Applied Chemistry–II

Applied Chemistry–II

V.M. Balsaraf (Sawargaonkar) M.Sc., Ph.D.
Datta Meghe College of Engineering
Airoli, Navi Mumbai-400708

Invited Postgraduate Lecturer
Department of Chemistry
University of Mumbai

TECHSAR™
NEW DELHI

Published and Printed by

Techssar Pvt. Ltd.
4435-36/7, Ansari Road, Daryaganj
New Delhi–110 002 (India)
E-mail: tech_info@techsarworld.com
Website: www.techsarworld.com

ISBN: 978-81-907770-9-4

Reprint 2025

PREFACE

I am very happy to present *Applied Chemistry-II* as per the revised syllabus of Mumbai University which is common to all branches of Engineering. The students of all academic backgrounds will find this book logical and easy to understand. This book presents applications of chemistry in various fields with the aim of providing clear and sufficient knowledge of chemistry to the students of engineering and technology. Each chapter has been dealt with in sufficient details and with different types of numericals. At the end of each chapter sufficient number of solved problems and quite a large number of university questions are provided. I hope that this will enable the students to have a complete understanding of the applied concepts of chemistry.

Various advanced technical developments like nanotechnology have proved that knowledge of chemistry is essential for better future of engineering and technologies. With this view all chapters in this book are written in simple language for the benefit of students. A laboratory manual is included for better understanding.

I thank my colleagues Ms. Elizabeth Thomas and Prof. G.B. Kunde for their continuous help in making this book more perfect. I also thank Dr. J.J. Kadam, Dr. S.D. Shete, Dr. S.N. Singh, Dr. S.N. Barai (Principal Tasgaonkar College), Prof Pujari, S.M. Patil, Anjali, Kirkire, Satpute and other teachers for their continuous encouragement.

Constructive suggestions/criticism from the reader/teachers is welcome for making this book still better in the future editions.

V.M. Balsaraf

CONTENTS

SYLLABUS

University of Mumbai

CLASS: F.E. (All Branches of Engineering)		Semester-II	
SUBJECT: Applied Chemistry–II			
Periods per week	Lecture	3	
01 Period of 60 min.	Practical	1	
	Tutorial	–	
		Hours	Marks
Evaluation system	Theory examination	2	75
	Practical	—	—
	Oral examination	—	—
	Term work	–	25
	Total	–	100

Detail of the syllabus:

Sr. No.	Details	Hrs
Module 01	Corrosion: • Nernst theory, Standard electrode potential. Types of corrosion dry or chemical corrosion wet or electrochemical corrosion. Electrochemical, Galvanic cell. Concentration cell. Intergranular stress cell corrosion. Polarization. Over voltage. Factors affecting rate of corrosion. • Methods to decrease the rate of corrosion. Cathodic and Anodic protection. Cathodic and Anodic coatings. Advanced coatings and Protection methods. Only constituents and their function of: (a) Paints (b) Varnishes (c) Lacquors (d) Enamels. • Metallic Coatings: Methods of coating and study only electroplating method. Corrosion Engineering of electronic and photonic devices.	(8)
Module 02	Alloys: • Alloys, Types of alloys, Alloys of Al, Cu and Pb. Their composition properties and used. Recent advances in alloy related materials. • Powder Metallurgy. Methods of metal powder formation. Metal ceramic powders. Technology of powder metallurgy applications of powder metallurgy.	(6)
Module 03	Fuel: • Definition. Classification.Characteristic properties of a good fuel. Calorific value. Cross and Net calorific value. Conversion Peoximate and altimate analysis of fuels. Combustion calculation for requirement of oxygen and air for given solid, liquid gaseous fuel. • Liquid fuels: Crude Petroleum oil. Classification. Separation and purification of Gasoline from crude oil. Thermal cracking. Catalytic cracking. Fixed bed, moving method for obtaining gasoline. • Diesel, Bio diesel, Methods to obtain bio diesel, Production of ethanol using bio-mass. Production of hydrocarbons from plants. Knocking, Octane value, Cetane Value, Antiknocking agents and their function, Recent technology for catalytic converter.	(10)

Module 04	Composite Materials: • Introduction, Constitution, Characteristic properties classification. Particle, fiber, reinforced composites structural composites. • Application of composite materials.	(5)
Module 05	Green Chemistry: • Introduction, Goals Significance, Basic ideas in the field of green chemistry research with 3 examples. • Industrial applications of green chemistry.	(5)
Module 06	Catalysis: • Introduction, Importance of catalysts and adsorbents in industry, Activation energy and catalysts. • Molecular design for catalysts and adsorbents, Molecular design by nature-zeolites, zeotypes, pillard clays, Metal complexes and clusters, Oxide materials carbon materials, membranes.	(6)

Theory Examination:

1. Question paper will comprise of total 7 questions, each of 15 marks.
2. Only five questions need to be solved.
3. Question 1 will be compulsory and based on entire syllabus.
4. Remaining questions will be mixed in nature (for example suppose Q. 2 has part (a) from module 3 then part (b) will be from any module other than module 3)

In question paper weightage of each module will be proportional to number of respective lecture hours as mentioned in the syllabus.

Term Work:

Term work shall consist of minimum five experiments and a written test. The contribution of marks for term work shall be as follows:

Laboratory work (Experiments and journal)	:	10 marks
Test (at least one)	:	10 marks
Attendance (Practical and Theory)	:	05 marks
Total	:	25 marks

The main certification and acceptance of TW ensures the satisfactory performance of laboratory work and minimum passing in the TW.

List of Expts:

1. Estimation of Cu iodometrically
2. Estimation of Zn complexometric titration
3. Estimation of Ni complexometric titration
4. Estimation of Al complexometric titration
5. Calorific value of solid or liquid fuel using Bomb calorimeter.
6. Preparation of membranes for filter any one. Demon.
7. CO_2 from air by orsats method.
8. Estimation of Fe by gravimetric method.
9. Estimation of Ni by gravimetric method.
10. Estimation of Sn iodimetrically.
11. Preparation of biodiesel from edible oil.
12. Synthesis of simple layered materials and their characterization.
13. Preparing simple composites and their characterization.

CHAPTER 1

CORROSION

1.1 INTRODUCTION

Metals when exposed to atmosphere undergo destruction or deterioration due to chemical or electrochemical impact of the environment on the surface of metals. It is a slow process, but losses incurred due to corrosion are of high order. The losses cannot be estimated by considering only the metal loss. The corrosion of metal objects like equipment, instruments, chemical plants, structure, etc. makes them inefficient by damage. In such cases, the indirect losses may be much higher than the direct losses. When a structure like building or bridge collapses due to corrosion problems, the loss also includes the loss of human life, properties, cost of repairing, etc.

Corrosion once thought a problem concerned for only metallurgist and chemist but now it is a field of universal study because of use of metals in various sectors may be in small or huge quantity. Small metal objects like pin or a huge metallic structure like bridge, corrosion impairs their proper function and use. Therefore, corrosion has been correctly referred to as 'cancer' of metals. The knowledge of corrosion is therefore of great importance in designing, construction and for other engineering purposes. Best designs and longer life constructions can be only achieved by the proper knowledge of corrosion mechanisms and control.

Corrosion is defined as the destruction of solid metallic material by chemical or electrochemical attack by the environment. Well-known examples are:

- Rusting of iron—reddish brown scale formation on iron and steel objects.
- Formation of green scales on copper vessels. It is due to the formation of basic cupric carbonate ($CuCO_3$ to $Cu(OH)_2$).

Why do metals undergo corrosion: Most of the metals except noble metals undergo corrosion by direct chemical attack or by electrochemical attack on the metal by corrosive environment.

1.2 DIRECT CHEMICAL CORROSION OR DRY CORROSION

It is corrosion which occurs when a metal is in contact with gases in the atmosphere. The atmospheric gases like oxygen, chlorine, hydrogen sulphide, sulphur dioxide and other anhydrous substances can react with the metals to form solid films of corrosion products of different kinds. The extent of such corrosions depends on (1) the chemical affinity between the metal and the substances in the environment, and (2) the nature of the corrosion product formed on the surface.

1.2.1 Oxidation Corrosion

Corrosion due to the attack of oxygen is known as oxidation corrosion. Dry oxygen does not attack most of the metals at normal temperature to any appreciable extent. However, alkali metals like Na, K, Rb, etc. and alkaline earth metals like Be, Mg, Ca, Sr, etc. are quite readily oxidised at normal temperature. At high temperature, almost all metals except the noble metals i.e., Ag, Au, Pt are oxidised. The oxidation reaction can be represented as below. In the case of a divalent metal,

$$M \longrightarrow M^{2+} + 2e$$

$$2e + \frac{1}{2}O_2 \longrightarrow O^{-2}$$

$$M^{2+} + O^{-2} \longrightarrow MO \text{ [Metallic oxide]}$$

When the surface of the metal exposed to atmosphere the oxidation corrosion occurs, the oxide layer produced forms a barrier between the metal and the environment. For the continuation of oxidation, either the metal ions should diffuse through the layer and come to the surface or oxygen must diffuse into the metal. As the metallic ions are normally smaller than oxide ions, they diffuse through the oxide layer and come near the surface.

The process of the oxidation of the metal is dependent on the type of the oxide film produced on the surface of the metal.

Stable protective layers (Non-porous)

These oxide layers are strongly adherent, non-porous and fine-grained in texture. Heavy metals like copper, tin, lead, etc. produce such protective layers.

$$2Cu + O_2 \longrightarrow 2CuO \text{ (Stable non-porous)}$$

$$4Al + 3O_2 \longrightarrow 2Al_2O_3 \text{ (Stable non-porpous)}$$

Unstable non-protective layers

Silver, gold, platinum form oxides when exposed to atmosphere which decompose back into metal and oxygen. Thus, these metals do not undergo oxidation/corrosion.

$$2Ag + O_2 \longrightarrow 2AgO \text{ (Unstable)} \longrightarrow 2Ag + O_2$$

Volatile non-protective layers

The oxide layer formed gets volatilised and every time fresh metal surface will be available for further attack. This causes continuous and rapid corrosion. Eg. Molybdenum oxide layer.

$$Mo + O_2 \longrightarrow MoO_2 \uparrow \text{ volatile}$$

Porous non-protective layers

If the oxide layer formed being porous, oxygen diffuses through pores and oxidation continues till the metal is completely converted into oxide.

$$4Fe + 3O_2 \longrightarrow 2Fe_2O_3$$

Stable porous iron oxide does not gives any protection to surface of metal hence corrosion proceeds.

1.2.2 Corrosion by Other Gases

Gases like H_2, SO_2, CO_2, Cl_2, H_2S, F_2, etc. can react with metals forming corresponding corrosion products and the extent of corrosion depends on chemical affinity between the metal and the gas in the environment. It also depends on the formation of protective and non-protective layer on the surface of the metal. When the corrosion film formed is adherent and non-porous, the extent of corrosion is reduced as the film acts like a barrier between the metal and the corrosive substance. Example: Action of Cl_2 on Ag forming AgCl film. When the film formed is porous or non-protective, the surface of the metal is slowly destroyed. Action of dry chlorine gas on tin to form stannic chloride which being volatile, leaves a fresh surface to the action of the environment and corrosion proceeds.

$$Sn + 2Cl_2 \longrightarrow SnCl_4 \uparrow$$

1.3 ELECTROCHEMICAL PRINCIPLES GOVERNING CORROSION

Electrode Potential

According to Nernest theory all metals have a natural tendency to pass into solution slowly producing metallic ions and releasing electrons which cause a negative charge to be developed on the metallic surface. There is also reverse tendency of metallic ions to accept electrons and form metal atoms. A dynamic equilibrium is established between the two processes which results in the development of a potential at the surface of the metal piece in contact with the solution. This potential is known as *electrode potential.*

Standard electrode potential or standard oxidation potential

The potential of an electrode depends on the nature of the metal, concentration of the ions and the temperature. The electrode potentials are determined in comparison with standard hydrogen electrode whose potential is arbitrarily taken as zero. The standard electrode potentials of different metals are determined by connecting the standard hydrogen electrode with metal electrodes in contact with a solution of its ions of unit activity.

Electrochemical series

The elements are arranged in the order to their standard electrode potentials.

Standard electrode potentials at 25°C.

Metal ion electrode	Potential in volts
1. Li/Li^+	−3.045 Reactive (Anodic metals)
2. K/K^+	−2.925
3. Ca/Ca^{++}	−2.866
4. Na/Na^+	−2.714
5. Mg/Mg^{++}	−2.363
6. Al/Al^{+++}	−1.662
7. Zn/Zn^{++}	−0.763
8. Cr/Cr^{+++}	−0.744
9. Ni/Ni^{++}	−0.23
10. Sn/Sn^{++}	−0.136
11. Fe/Fe^{+++}	−0.045
12. $H_2/2H^+$	−0.000 References
13. Cu/Cu^{++}	+0.337
14. Ag/Ag^+	+0.80
15. Pt/Pt^{++}	+1.2
16. Au/Au^{+++}	+1.42 Noble (Cathodic)

Electrochemical series shows the relative tendencies of the metals to go into solution. Those metals having higher negative electrode potential displaces hydrogen from acid solutions. The metals higher in the electrochemical series can displace the metals lower in series from their solution.

Galvanic series

Electrochemical series is useful in understanding the chemical reactivity of the various metals. Electrode potentials of certain very common metal and alloys have been studied using calomel electrode as the common electrode in presence of sea water. The electrode potentials of the metals and alloys are arranged in the order of their activity and this arrangement is called 'galvanic series'. This series gives a better information regarding the relative corrosion tendencies of metals and alloys. The extent of corrosion is dependent on the difference in potential between the anode metal and cathode metal or alloy. The position of the metal or alloy is still influenced by certain interfering factors.

Galvanic series

1.	Magnesium (more anodic)	11.	Bronze
2.	Magnesium alloys ↑	12.	Copper Nickel alloys
3.	Aluminium	13.	Silver
4.	Aluminium alloys	14.	Stainless steel (passive)
5.	Low carbon steels	15.	Monel
6.	Stainless steel (active)	16.	Titanium
7.	Lead tin alloys	17.	Gold ↓
8.	Lead	18.	Platinum (more cathodic)
9.	Brass		
10.	Copper		

1.4 ELECTROCHEMICAL OR WET CORROSION

Corrosion of metals is caused due to formation of electrochemical cells on the surface of the metals. This type of corrosion occurs at the solid-liquid interface when the metals remain in contact with moist air or any other liquid medium. The movement of the electrons in the solid state takes place through a metallic conductor and in solution as ions. This type of corrosion involves the formation of galvanic or concentration cells of different magnitudes depending on the nature of the metal and the surroundings. The corrosion thus occurs due to the formation of anodic and cathodic areas through which current flows in presence of a conducting liquid. The anode undergoes oxidation or dissolution producing the electrons which are then consumed at the cathode where reduction occurs. The dissolved substances in the medium may absorb electrons to form ions like OH^- or O^{2-}. The metallic ions produced at the anode and the ions produced at the cathode part diffuse through the medium to form corrosion products somewhere between the anode and the cathode. The continuation of the electrochemical process is dependent on the nature of the product produced.

- If a soluble product or a gaseous product is produced, the corrosion of the metal continues.
- If the product formed deposits as an insoluble protective layer on the anode, corrosion is reduced.

1.4.1 Corrosion Due to Formation of Galvanic Cells

Galvanic corrosion

When two dissimilar metals of different electrode potentials are in contact with each other in the presence of an electrolyte or even when moisture formation of galvanic cell takes place. A metal when immersed in a solution acquires a potential depending on the nature of the metal and solution. The more active metal undergoes oxidation and less active causes reduction. The oxidations and reductions occur at the anode and the cathode and the galvanic cell formed develops an emf crossing the dissolution (corrosion) of the anode. The emf generated is called the corrosion current, if the emf generated is larger, the corrosion is faster.

When two metals of different electrode potentials are in contact with each other in the presence of an electrolyte.

- The metal which occupies a higher position in the electrochemical series behaves as the anode and gets corroded e.g. Zn, Mg, Al.
- If the difference in the electrode potentials of the two metals is more, the rate of corrosion is faster.
- When the metals forming the galvanic cell have almost similar electrode potentials, the tendency of the anodic metal to undergo corrosion is least.
- The rate of galvanic corrosion is also dependent on the ratio of the cathode to anode areas. Corrosion is found to be severe when the ratio of the cathode to the anode area is large.

Instance of galvanic corrosion:

- Use of steel pipes in copper storage tanks.
- Soldering of copper wires with lead-antimony solders.
- Use of steel screws in any equipment made of brass.

1.4.2 Corrosion Due to Formation of Concentration Cells

Electrochemical corrosion also occurs due to the formation of concentration cells. Concentration cells are developed either due to variations in the concentrations of the metal ions in contact with a metal or due to difference in the availability of air or oxygen over the metal surface which sets up a difference in potentials at different places.

Metal ion concentration cells

Electrode potential of a metal depends on the concentration of its own ions in the solution. The electrode in contact with a dilute solution has a preferential tendency to dissolve and become the anode while that in contact with a more concentrated solution behaves cathodic.

Same metal electrodes in contact with solutions of different concentrations of metallic ions at different parts of the metal piece is subjected to corrosion due to formation of metal ion concentration cells. The part of the metal piece which is in contact with the solution of lesser concentration of ions behaves anodic and gets corroded whereas the other part is saved from corrosion. The rate of corrosion depends on the difference in the concentration of the metallic ions at the two places. There is severe dissolution of the anodic part when there is a large difference in the metal ion concentrations. E.g. Cu electrodes dipped in a $CuSO_4$ solution of different concentrations, Zn electrodes dipped in a $ZnSO_4$ solution of different electrodes, etc.

Oxygen concentration cells (Differential aeration corrosion)

The difference in the availability of air or oxygen over a metallic surface leads to formation of oxygen concentration cells. The less aerated or less oxygenated part behaves anodic while the more oxygenated part cathodic. Since cathodic reactions involve consumption of oxygen, the more oxygenated part behaves cathodic and less oxygenated part behaves anodic. The corrosions which are caused due to formation of such oxygen concentration cells are known as differential aeration corrosions. It is believed that aeration causes the formation of an oxide film on the metal surface resulting in the development of a metal metal-oxide cell or even oxygen electrode-metal electrode system.

Most of the common types of corrosion are due to differential aeration. The following are some of the examples:

(a) Corrosion of metals partly immersed in solutions

When a metal piece is partly immersed in a solution or water, the part of the metal piece above the surface of the solution is more aerated and behaves cathodic while the parts immersed in interior has less accessibility to oxygen and behaves anodic. The difference of potential created causes the flow of current between the two regions which results in corrosion.

Anode $$Fe \longrightarrow Fe^{2+} + 2e^-$$

The electrons liberated are consumed by oxygen at the cathode to form hydroxyl ions.

$$\frac{1}{2}O_2 + H_2O + 2e^- \longrightarrow 2(OH)^-$$

This results in the corrosion or rust formation in the anodic region. The reaction is as below:

$$Fe^{2+} + 2(OH)^- \longrightarrow Fe(OH)_2\downarrow$$

$$Fe(OH)_2\downarrow \xrightarrow{O_2} \underset{\text{rust}}{Fe(OH)_3}\downarrow$$

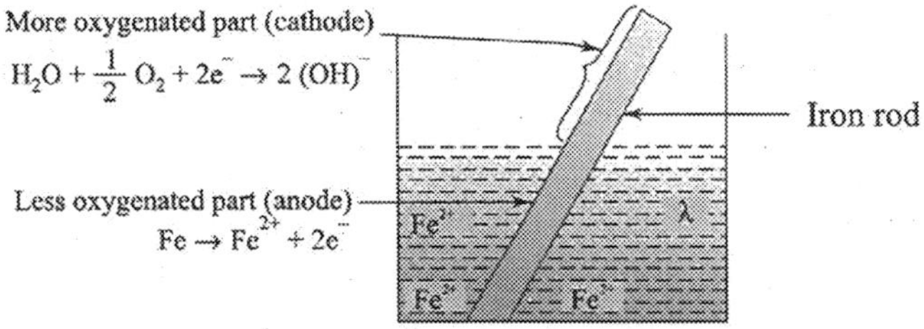

Fig. 1.1 Corrosion of partly immersed iron in water.

(b) Corrosion under a drop of a solution

Metals like iron undergo corrosion below a drop of a conducting solution or even water due to formation of concentration cells on account of differential aeration.

The part of the metal surface covered by the drop has less accessibility to oxygen and behaves anodic (less oxygenated metal part) while the other part having more accessibility to oxygen behaves cathodic (more oxygenated metal part).

At the centre of the drop,

$$Fe \longrightarrow Fe^{2+} + 2e^-$$

At the edges of the drop,

$$\frac{1}{2}O_2 + H_2O + 2e^- \longrightarrow 2(OH)^-$$

Near to the centre of the drop

$$Fe(OH)_2 \downarrow \xrightarrow{O_2} \underset{\text{rust}}{Fe(OH)_3} \downarrow$$

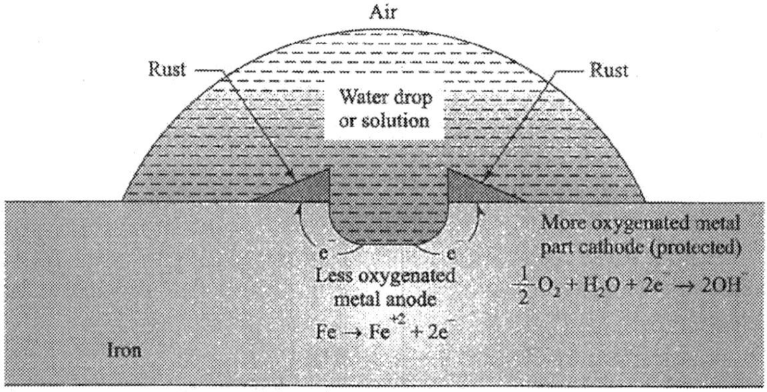

Fig. 1.2 Corrosion of iron surface covered by the drop of water or solution.

(c) Corrosion occurring cracks on metal surfaces or areas where accumulation of dust, rust, sand, etc. occurs

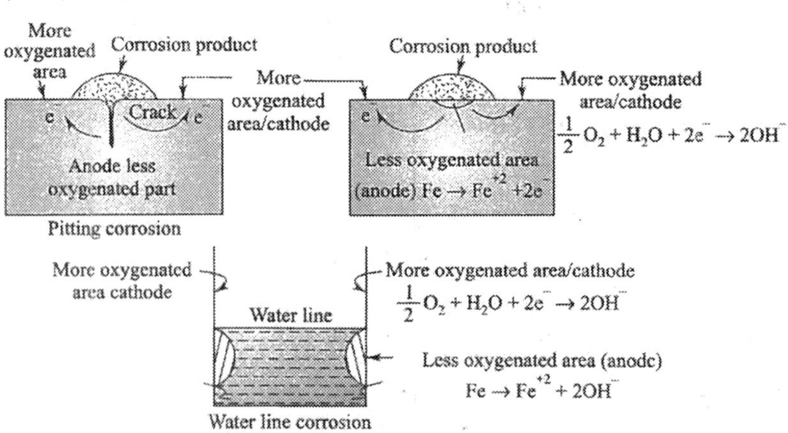

Fig. 1.3 Examples of corrosion due to difference of oxygen concentration in different parts of metal.

All the above examples of corrosion illustrate that the corrosion caused due to differential aeration results in a localised attack on oxygen deficient parts causing pitting on surfaces. The attack becomes severe slowly due to the accumulation of the corrosion product on the affected region, thereby screening it further from oxygen or air which results in an increased potential difference.

1.4.3 Mechanism of Electrochemical Corrosion

There are mainly two types of corrosion based on the reactions involved in the process. (a) Hydrogen evolution type (b) Oxygen absorption type.

Corrosion of a metal is electrochemical in nature involving the movement of the electrons between the anodic and cathodic regions. The anode undergoes dissolution and some substance is liberated at the cathode.

At the anode, the metals lose their electrons and are oxidised to positive ions.

$$Fe \longrightarrow Fe^{2+} + 2e^-$$

The ferrous ions formed combine with hydroxyl ions in the solution forming less soluble $Fe(OH)_2$.

$$Fe^{++} + 2(OH)^- \longrightarrow Fe(OH)_2$$

$$Fe(OH)_2 \xrightarrow{O_2} \underset{\text{rust}}{Fe(OH)_3} \downarrow$$

The electrons released at the anode move towards the cathode where the following reactions occur.

At the cathode, reactions are different depending upon the availability of oxygen near the cathode.

Hydrogen evolution type

(a) If the medium is acidic,

$$2e^- + 2H^+ \longrightarrow H_2(g) \uparrow$$

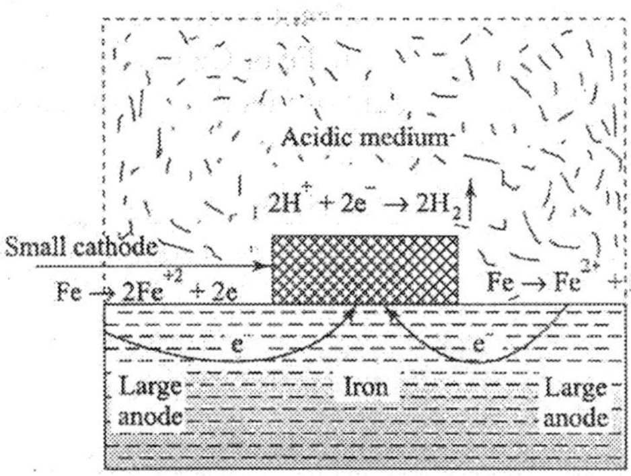

Fig. 1.4 Hydrogen evolution type.

Corrosion mechanisms which liberate hydrogen are called hydrogen type of corrosions.

Oxygen absorption type

(b) In alkaline or neutral medium, oxygen is absored to produce rust.

$$2e^- + H_2O + \frac{1}{2}O_2 \longrightarrow 2(OH)^-$$

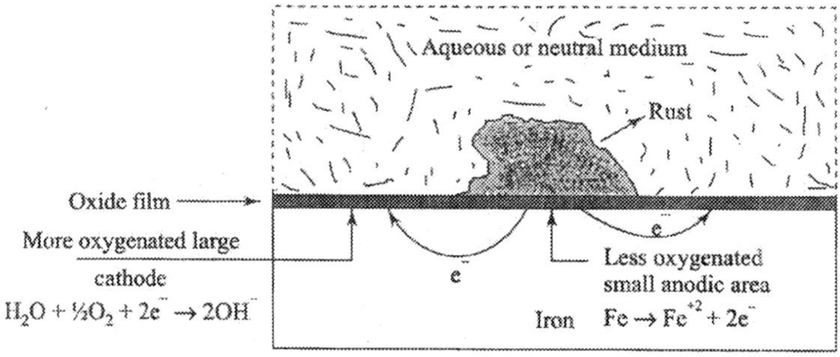

Fig. 1.5 Oxygen absorption type.

1.5 FACTORS INFLUENCING THE RATE OF CORROSION

(i) Electrode potential or position of metal in galvanic series

The more electropositive metals which have a higher negative electrode potential get corroded faster in the presence of less electropositive metal and the rate of corrosion increases with the difference in the potentials of the two metals.

(ii) Purity of the metal

Impurities in a metal cause a heterogeneity and thereby formation of separate anodic and cathodic area. Even traces of impurities lead to formation of electrochemical cells whereby a significant corrosion of the anodic metal takes place. Presence of a small amount of Pb, Fe or Cu in zinc causes severe corrosion of Zn. It is observed that small amount of metal impurity causes large amount of corrosion.

Metal Zn

% Impurity	Rate of corrosion
0.001	1
0.010	2.650
0.05	5000

(iii) Hydrogen overvoltage

The rate of corrosion of zinc in contact with 1 N H_2SO_4 is found to be lower as hydrogen is liberated in the form of bubbles on the metal surface, the overvoltage being 0.70 volts. Addition of a few drops of $CuSO_4$ will cause deposition of copper on Zn metal and the overvoltage is reduced to 0.33 volts. Thus, it is noticed that the rate of corrosion is accelerated by reducing the overvoltage of the metal concerned.

(iv) Physical state

The rate of corrosion is influenced by grain size and the orientation of the crystals on the surface. The regions of high residual stress tend to be anodic with respect to the stress free surface. This results in stress corrosion. An uneven surface also get corrodes due to the presence of crests and troughs.

(v) Passivity and nature of the oxide film

The rate of corrosion mainly depends on the nature of oxide films and also its reaction with the environment. The rate of corrosion is more whenever the volume of the oxide film formed is less than the metal as in the case of metals like Mg, Ba, Li, Na, etc. The metals like Al, Cr, W and Ni produce voluminous and protective oxide layers.

(vi) Nature of the corrosion product

Protective, non-protective or volatile layer may be formed by corrosion product. When the corrosion product formed exists as an insoluble, dense and adherent layer on the surface of the metal, the rate of corrosion decreases. If the product formed is insoluble in the environment, the rate of corrosion increases.

(vii) Relative sizes of the anodic and cathodic areas

If the anodic area is very small in comparison to cathodic area severe corrosion often occurs. Scratch on a tin coated steel sheet produces severe localised corrosion.

1.5.1 Environmental Factors which Affect Rate of Corrosion

Effect of pH

Acidic mediums are more corrosive as compared to neutral or alkaline solutions. Amphoteric metals like Al, Zn, Pb, etc. are stable in neutral media but dissolve in acid and alkaline solutions forming complex ions. The rate of corrosion of iron is found to be much slower in oxygen-free solutions and the rate increases with the decrease of pH. The rate of corrosion is very high at low pH, which can be reduced in several cases by increasing the pH of the medium. For example, boiler-feed water is kept in the alkaline range for saving the boiler metal from stress cracking and pitting.

Oxygen concentration and its influence

The rate of corrosion is influenced by the concentration of oxygen. Larger the difference in the concentration of oxygen, larger is the rate of corrosion on a surface. The presence of dissolved oxygen in boiler feed water results in different types of corrosion problems in boilers and heat exchangers. The limit of dissolved oxygen in high pressure boiler is 0.005 mgs/litre.

Presence of oxidising agents causes oxidation of the metals and the oxidising agents get reduced near the cathode promoting the cathodic reactions. Metals like Al, Mg and stainless steel form oxide films on the surface which make the metal resistant to corrosion.

Presence of cations and anions

Rate of corrosion may be influenced by the presence of chloride ions which destroy the passive film i.e. protective oxide films on stainless steel and aluminium. Silicate ions react to form silica gel which inhibits corrosion. The dissolved salts normally make the solution more conducive and thereby the rate of corrosion increases. Presence of inhibitors produces different effects. These are organic or inorganic substances. Substances like silicates, borates, chromates, etc. act as anodic inhibitors and suppress anodic reactions. Cathodic inhibitors form resistant oxide films or insoluble salts which form deposits on the metallic surfaces. Organic inhibitors which are colloidal in nature form protective layers on the metal surface. Organic substances like amines, pyridine, quinoline, etc. are used. When inhibitors are used they are used in sufficient quantities so that they form protective film over the entire region.

Effect of temperature

The rate of chemical reactions increases with rise in temperature up to a particular limit after that it remains constant or decreases but the solubility of oxygen decreases. Hence, the rate of corrosion due to the presence of oxygen and oxidising agents decreases with the rise in temperature whereas rate of corrosion in the absence of oxygen with rise in temperature increases. The rate of corrosion of mild steel increases rapidly up to 80°C beyond which the rate of corrosion decreases.

Humidity

The rate of corrosion of metals increases rapidly above the critical humidity. Critical humidity depends on the physical characteristics and the nature of the corrosion product. Increased rate of corrosion in humid atmosphere is due to the gases present in the atmosphere which get dissolved in moisture giving conducting solutions. Corrosion of buried pipelines under the mineralised soils is found to be more severe than those under sandy soils. The rate of corrosion also increases when the atmosphere contains chemically active substances which increase the conductivity in the presence of moisture.

1.6 TYPES OF CORROSION

1.6.1 Pitting Corrosion

This type of corrosion mainly occurs due to the presence of pits or cavity.

Pits are formed due to deposition of dust or dirt on the surface of the metal. This can also occur due to rupture of protective oxide film on the surface of the metal. The covered part is small as compared to the rest of the area. The covered area acts as anode and the cathode area made up of the rest of area is so large as to induce accelerated corrosion of anode. Uneven surface scratches, damage to surface due to abrasion, impact, etc. can lead to formation of areas susceptible to pitting corrosion.

Pitting corrosion is differential aeration type corrosion and can be explained by the formation of oxygen concentration cells. The part of metal surface covered with dust is less exposed to oxygen or air as compared to the rest of the area of the metal. The less oxygenated area below the dust acts as anode and undergoes corrosion.

Anode $$Fe \longrightarrow Fe^{2+} + 2e^-$$

In the presence of moiture containing dissolved oxygen, the electrons released due to dissolution are accepted at cathode.

$$H_2O + \frac{1}{2}O_2 + 2e^- \longrightarrow 2(OH)^-$$

$$Fe^{2+} + 2(OH)^- \longrightarrow \underset{\text{Corrosion product}}{Fe(OH)_2 \downarrow}$$

leading to the formation of $(OH)^-$ ions which combine with ferrous ions to produce corrosion product.

Due to the above reactions, the covered metal part gets more covered with the deposition of corrosion product causing the corrosion problem. Pits formed travel downward causing holes in the pipes and reaction vessels. When the fluids pass through pipes under pressure, areas around pits are weak and give way, leading to rupture of pipes.

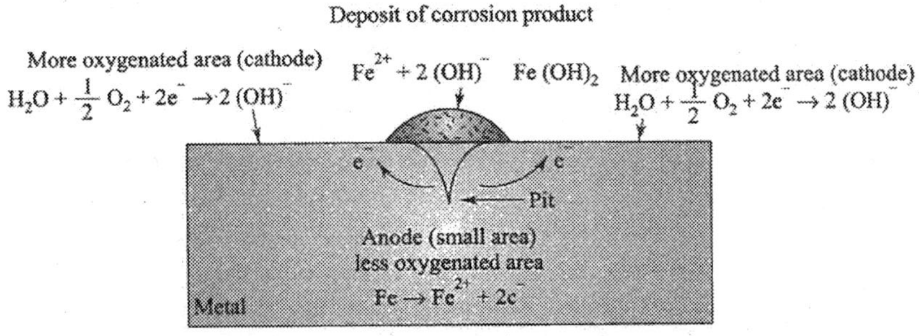

Fig. 1.6 Pitting corrosion at the surface of iron.

This type of corrosion can be minimised by ensuring that the deposit on the surface are not allowed to be formed and no damage to surface occurs. In holding vessels and pipes, deposit formation can be avoided by filtering the liquid so that no dust can settle down. Further in reaction and mixing vessels, adequate stirring arrangement is made so that there is no chance anything settling down to form a deposit. Further all parts of the vessel must be accessible so that they can be periodically inspected and cleared thoroughly.

1.6.2 Crevice Corrosion

Like pitting corrosion, intensive and localised corrosion occurs in shielded area and crevices. This type of corrosion is the result of stagnation of liquid in small holes, crevices, gasket surfaces, etc. Deposits of sand, dust and corrosion products can also initiate crevice corrosion. These deposits act as shield and create stagnant conditions. Where gaskets are used between two metallic surfaces, if the gaskets are not fitted properly, conditions may exist where small amounts of liquids are retained on metal surface. Fibrous gaskets can also retain liquid due to capillary action and results in collection of small pools of liquids.

Crevice corrosion is illustrated by considering corrosion occurring near rivet joint immersed in sea water. Corrosion is believed to take place either due to difference in metal ion concentration or oxygen concentration between the crevice and its immediate neighbourhood. The reactions are

At anode $M \longrightarrow M^+ + e^-$ (Oxidation)

At cathode $O_2 + 2H_2O + 4e^- \longrightarrow 4(OH)^-$ (Reduction)

Reduction reaction at the cathode proceeds as long as dissolved oxygen is available. The oxidation reaction will result in formation of metal ion in solution. In the absence of $(OH)^-$ ions to neutralise metal ions, chloride ions in sea water migrate towards metal ions and form metal chloride.

$$M^+ + Cl^- \longrightarrow MCl$$

Magnesium chloride and to lesser extent calcium chloride undergo hydrolysis at high temperatures.

$$MgCl_2 + 2H_2O \longrightarrow Mg(OH)_2\downarrow + 2HCl$$

This HCl reacts with metal forming metal chloride as

$$Fe + 2HCl \longrightarrow FeCl_2 + H_2\uparrow$$

The metal chloride formed again undergoes hydrolysis and produces HCl.

$$FeCl_2 + 2H_2O \longrightarrow Fe(OH)_2\downarrow + 2HCl$$

Because of liberation of HCl again and again corrosion becomes more and more intense.

Following procedure can be used to reduce crevice corrosion:

- As far as possible avoid riveted and bolted joints, instead weld.

- Close crevices in existing lap joints by continuous welding.
- Design vessels for complete drainage, avoid sharp corners and stagnant areas.
- Inspect equipment periodically and remove deposits.
- Remove suspended impurities which settle to form deposits early in the process by filtration.
- Use solid non-absorbent gaskets.

1.6.3 Intergranular Corrosion

During the casting of metal in molten condition, the grains are formed. The arrangement of the atoms and the spacing between the atomic layers are found to be same in all grains in a particular metal. However, the atomic planes of adjoining grains do not match. Such regions between the grains are known as grain boundaries. These boundaries are more susceptible to the attack by a corrosion substance in the environment. The type of corrosion that occurs along the boundaries of the grains is termed intergranular corrosion.

Intergranular corrosion is often due to the presence of certain precipitated substance close to the grain boundaries which show more anodic potential compared to the substance at the grain centre. The precipitation of certain compounds at the grain boundary leaves the solid solution adjacent to the grain boundary depleted in one of the essential constituents that may be needed for corrosion resistance. Since this grain boundary is anodic with respect to the grain centre as well as to the precipitated substance, it is preferentially attacked by the corrosion environment. The corrosion occurs along the path of the grain boundaries on a microscopic scale producing weakness and brittleness in the metal. This results in sudden failure of the material due to loss of cohesion between the grains and it occurs often without any apparent indication of localised electrochemical attack.

This type of corrosion is observed commonly in the alloys of Al, Cu and welded stainless steel (containing Fe, Cr, C), chromium carbide is precipitated at the grain boundaries which causes depletion of chromium in adjacent regions of the grain boundaries thereby establishing anodic regions of higher electrode potential. Thus, the regions near the grain boundaries become more susceptible to corrosion. This corrosion in welded stainless steals and other alloys can be prevented by proper heat treatment followed by rapid quenching to avoid the heterogeneous precipitation of substances that may result in slow cooling.

1.6.4 Stress Corrosion

This type of corrosion is characterised by formation of fine hair cracks in areas subjected to stress in boilers and cartridges.

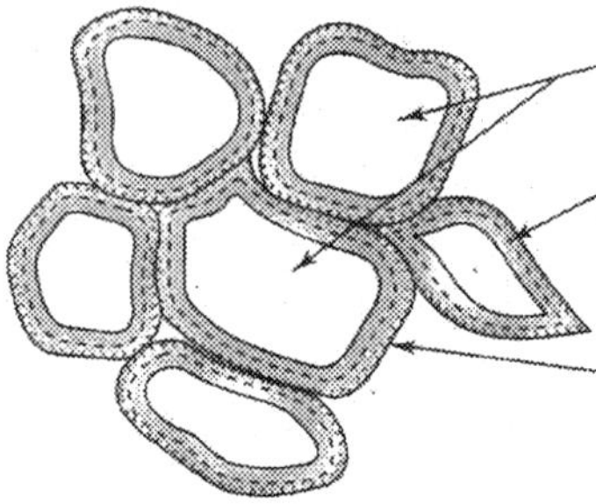

Fig. 1.7 Intergranular corrosion.

In boiler plates near the riveted joints hair cracks are formed due to stress. Whenever explosions in boilers were investigated, it was found that these hair cracks were filled with NaOH and iron reacted with NaOH to form sodium ferrate. Hence, these areas became weak and yielded to high pressure of boiler leading to explosion. The source of NaOH is Na_2CO_3 used for softening of water. Na_2CO_3 undergoes hydrolysis at high temperature of boiler to produce NaOH and CO_2.

$$Na_2CO_3 \xrightarrow{H_2O} 2NaOH + CO_2 \uparrow$$

This stress corrosion in boilers is also explained on the basis of formation of electrochemical concentration cell as follows:

$$\underset{\text{near rivets}}{\overset{\oplus}{\text{Iron}}} \left| \text{ Con. NaOH } \right\| \text{ Dilute NaOH } \left| \underset{\text{rest of material}}{\overset{\ominus}{\text{Iron}}} \right.$$

Iron exposed to higher concentration of NaOH behaved as anode and the rest of iron plate in contact with dilute NaOH behaves as cathode, setting up a concentration cell where the anode undergoes corrosion, known as caustic embrittlement.

During rainy season the region where brass cartridge is crimped to bullet undergo cracking in presence of amonia. Similarly, stainless steel develops cracks when exposed to chloride ions over a period of time, whereas stainless steel is not affected by ammonia.

Stress corrosion is usually traced to materials undergoing some mechanical treatment like rolling, drawing, bending, etc. where the materials have not been sufficiently annealed. Further stress may be present initially or may develop due to service conditions. The atoms of metal under stress are at higher energy level than the surrounding areas and behave as anode and undergo corrosion.

This type of corrosion can be considerably reduced by ensuring that sections are subjected to stress relief annealing. During the installation stage sufficient precautions can be taken to see that no part of structure is under stress.

1.6.5 Biological Corrosion

Corrosion is influenced due to the presence of microorganisms. Microorganisms have been found to reproduce even in adverse conditions like high acidity, high temperature and at high pressures. Hence, under normal conditions they can grow fast in the presence of sufficient nutrients in the vicinity. Broadly organisms are classified as anaerobic organisms and aerobic organisms. In the process of growth and reproduction, microorganisms utilise the nutrients present for their metabolic activities which influence corrosion process as follows:

- Produce deposits which may create conditions for corrosion.
- Produce products which can have corrosive action.
- Produce products which destroy the protective film on the surface.
- Interfere with anodic and cathodic reactions.

The population of microorganisms are changing depending upon various factors in soils. Dry weather leads to better access to oxygen and hence there is a preponderance aerobic microorganisms. In rainy season, on the other hand, due to waterlogging, conditions are more favourable for the growth of anaerobic microorganisms. In winter months when the temperatures is lower only those organisms which can survive cold and grow under those conditions, will grow. The other organisms will not grow at all. These seasonal changes will change the environment due to difference in the metabolic activities of organisms.

Examples of corrosion due to microbial activity are given below.

Corrosion due to sulphate reducing bacteria

Sporovibrio desulphuricans organisms which grow under anaerobic conditions cause extensive damage to buried steel structures. These organisms can grow well in temperature range 25–30°C and under pH range 5.0 to 9.0 and in the presence of sulphate in soil. The corrosion reactions

$$Fe^{2+} + H_2S \longrightarrow FeS + 2H^+$$

$$3Fe^{2+} + 6OH^- \longrightarrow 3Fe(OH)_2$$

The organisms bring about oxidation of iron through complex set of reactions and suggested route as follows:

$$8H_2O \longrightarrow 8H^+ + 8OH^-$$

$$4Fe + 8H^+ \longrightarrow 4Fe^{2+} + 8H$$

and $$H_2SO_4 + 8H^+ \longrightarrow H_2S + 4H_2O$$

Usually, buried structures show deposits of black iron sulphide and brown ferrous hydroxides, which cause other types of corrosion. Under aerobic conditions another set of bacteria belonging thiobacillus group oxidise sulphur to sulphuric acid which lower the pH of surrounding and thereby cause corrosion.

Iron and manganese bacteria are other group of aerobic bacteria which have been found to cause corrosion of iron and steel structures.

This corrosion can be minimised by preventing microbial growth or by ensuring that conditions do not favour growth of organisms or by including agents which will prevent the growth of organisms.

1.7 CORROSION CONTROL OR PROTECTION FROM CORROSION

In India a large amount of metal is destroyed every year due to corrosion. It is necessary to give protection to metals from corrosion by controlling the corrosion in a suitable manner. It is difficult to control corrosion completely but it can be minimised to an extent. The method adopted depends on the concerned metal or object and the environmental conditions to which the metal is exposed. Some of the important methods of controlling corrosion are given below.

1.7.1 Control of the Environmental Conditions

- Elimination of corrosive substances: may be done by avoiding contact or addition of additives.
- De-oxygenation: chemical treatment by addition of sodium sulphite, sodium sulphide, hydrazine, etc.
- De-humidification: elimination of moisture can be obtained by using substances like silica gel, alumina, etc.
- Corrosion incubators: by using cathodic and anodic inhibitors.

1.7.2 Proper Desigining and Material Selection

Selection of proper material and proper designing for construction can minimise corrosion to a significant extent. Noble metals are resistant to corrosion but their applications are limited as they are costly.

- Metal when used for fabrication should not contain any cathodic impurity. For example: Presence of iron in pure Mg in quantities more than 0.003% appears as small particles and leads to pitting.
- As far as possible, when two metals are used, the metals chosen should be as close as possible in electrochemical series. The anodic metal in the combination need not be painted as any break in the coating will cause severe localised corrosion.
- Proper heat treatment, proper metals, alloys may be used by giving so that internal stresses are eliminated which will improve corrosion resistance.
- Design should be such that as far as possible joints, bends, sharp corners to be avoided so that accumulation of dust, moisture would not take place but free circulation of air is possible.

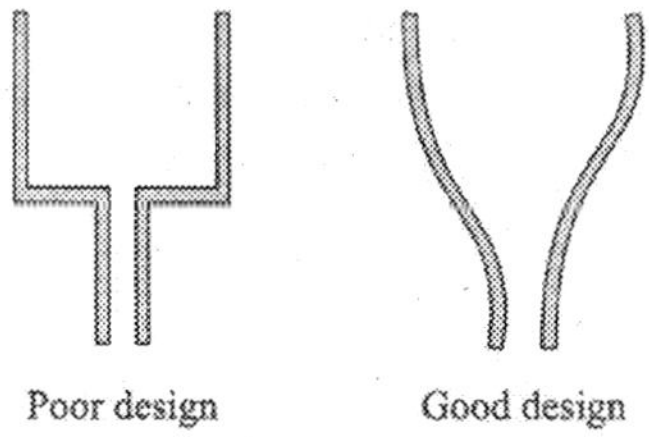

Fig. 1.8

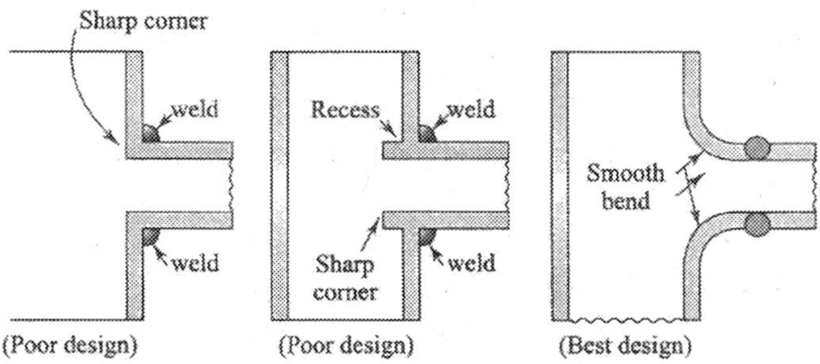

Fig. 1.9 Prevention of corrosion by design—elimination of sharp corners and stress.

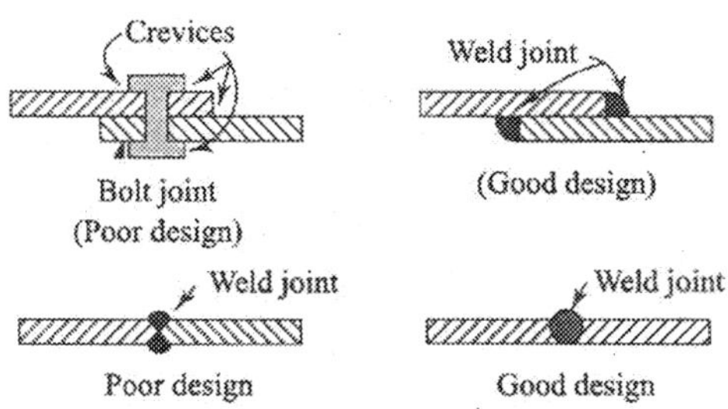

Fig. 1.10 Light Joints.

1.7.3 Cathodic Protection

Corrosion can be controlled by this method when it is not possible to alter the environment of the concerned metal. When anodic and cathodic areas are developed on the corroding surface, the corrosion of the anode will be faster when the current produced is larger. In cathodic protection, an opposing current is applied to nullify the current generated in the system.

In this method, corrosion is reduced by forcing the metal to behave like cathode in the corrosion environment.

By applying impressed current

In this method, an external emf from a source like battery or rectifier is applied to the corroding system in such a manner that the article is forced to behave like cathode which gets protected. The anode is either an inert material like graphite or one which undergoes destruction and needs to be replaced. Other anode materials in use are high silica iron, carbon, stainless steel and platinum. The anodes are buried in backfill such as coke, breeze or gypsum. This method is useful in giving protection to buried items such as tanks, pipelines, transmission line towers, condensers, etc.

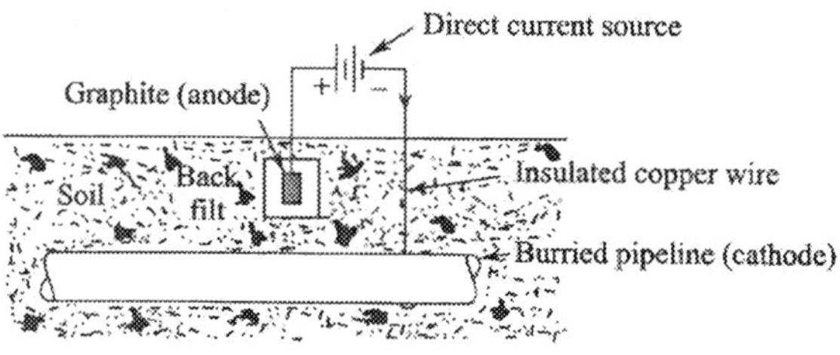

Fig. 1.11 Impressed current cathodic protection.

Drawbacks

- This method requires high current which increases maintenance costs.
- The current may not be uniform over the entire surface of the metal.
- Sometimes hydrogen gas is liberated at cathode and this may result in hydrogen embrittlement.

By using sacrificial anode

In this method, a more active metal like Zn, Mg, Al, etc. is connected to the material to be protected. The more active metal behaves anodic thereby undergoing destruction. These anodes are called sacrificial anodes and they can be replaced when corroded. This method is used for protecting cables, pipelines and other burried structures from soil corrosion. This method is also applicable for saving ship hulls, cables, marine structures, etc. from salt water corrosion. Magnesium rods or pieces are inserted into domestic water boilers or tanks for preventing corrosion.

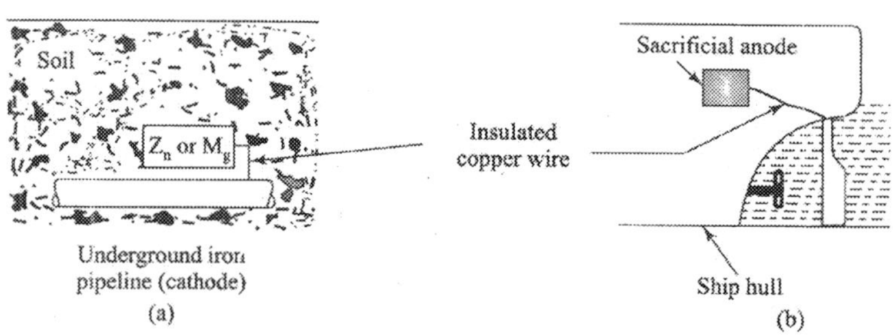

Fig. 1.12 Sacrifical anodic protection.

1.7.4 Application of Protective Coatings

Protective coatings provide a satisfactory barrier between the metal and the environment and thereby corrosion reactions do not occur. Before applying any kind of protective coating, surfaces should be cleaned properly. Different types of cleaning procedures are adopted for cleaning depending upon the type of the equipment and the nature of the environment.

The different types of protective coatings include metallic, inorganic and organic coatings.

- Metallic coating, e.g. hot dipping, metal spraying, metal cladding, cementation and electroplating
- Inorganic coatings, e.g. chromate, phosphate, oxide coating and anodisation
- Organic coating, e.g. paints, varnishes, enamels, lacquers.

1.7.5 Preliminary Treatment of Surfaces (Cleaning of the Article to be Coated)

In order to achieve a strong adhering and durable coating or to secure long life service of the coating the surface of metal should be properly cleaned. As the metal surface is exposed to the atmosphere, it will have a thin layer of oxide film impurities, rust, scales oils, greases, etc. Unless these impurities are completely removed from the surface, the coating will not adhere properly to the surface and durability of such coatings will also be poor. Some of the general methods of surface preparations are given below:

Degreasing

Oil and greases may be resumed by cleaning with organic solvents like CCl_4, $CHCl_3$, acetone. Immersion in hot alkali saponifies the fatty material which can be easily removed by washing with hot water. To remove old paint layer, a mixture of trisodium phosphate alkali and soap is used. After alkali treatment, the surface is thoroughly washed with water and to remove the last traces of alkali, it is washed with 0.1% chromic acid. Finally, the surface is again treated with fresh water.

When the surface is less contaminated, washing with steam and hot water, will be sufficient to remove the surface impurities.

Acid treatment (descaling)

Acid treatment is also known as pickling process. This involves immersing the material to be coated in an acid bath. The concentration of acid solution and duration of immersion, temperature of bath, vary depending upon the thickness of oxide coating and process requirements. After the acid treatment, the surface is completely washed with water.

Mechanical cleaning

The loose deposits on the surface can be removed by wire brush. Sand blasting is also used for treating large surface areas as storage tanks, etc. Sand blasting involves directing a blast of fine sand against the surface by means of compressed air. This dislodges the scales and removes the oxide layer from the surface.

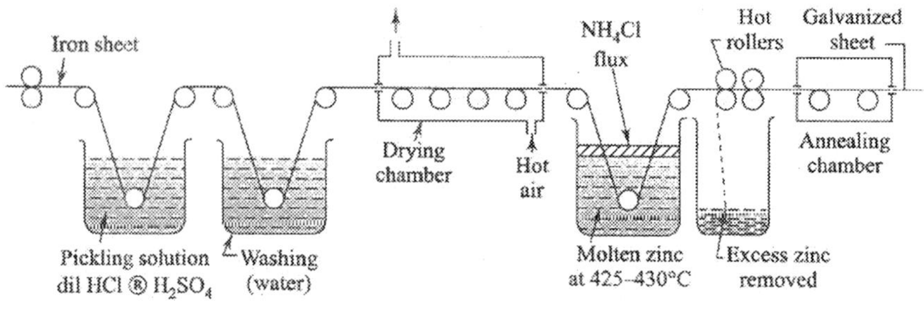

Fig. 1.13 Galvanising.

Electrochemical method

It is used to remove oxide scales which are not removed by other method. The base metal is made either anode or cathode with an electrolyte acid or base. At anode oxide scale dissolves in electrolyte leaving behind base metal. At cathode the metal oxides are reduced to metal.

1.7.6 Metallic Coatings

They are widely used on steel and iron which are used for various construction purposes. Coatings by metals like Zn, Sn, Ni, Cu, Cr, Al and Pb are used. The anodic coatings of Zn, Mg, Al, Cd are found to be superior to cathodic coatings of Sn, Cu, Ni, Cr, etc. on iron.

Methods of Applications of Metallic Coatings

1.7.6.1 Hot cipping

In this method of metallic coating, the article to be coated is dipped in a molten bath of coating metal. Galvanising and tinning are two examples of hot dipping.

(a) Galvanising (Anodic coating): The following diagram illustrates the various stages in galvanising process.

Generally, low carbon steels are galvanised to protect them from corrosion. Mild steel sheets are galvanised in the following manner. The sheet metal is passed through a pickling tank containing dilute HCl or H_2SO_4 solution.

After pickling the sheet is thoroughly washed by passing it through another tank containing fresh water.

The sheet is then dried by passing through drying chamber, heated by passing hot air. The sheet then passes through bath of molten zinc. The surface of the molten metal is covered with a layer of ammonium chloride to avoid oxidation of molten metal. The sheet leaving the zinc bath gets coated with a thin layer of zinc metal. The surface of the metal initially forms an alloy with zinc, the zinc metal then gets deposited on the alloy.

The coated sheet passes through a set of hot rollers where excess of zinc deposited is removed.

The sheet then passes through annealing chamber so that the coating becomes firmly attached to the surface of the metal. The coated sheet is then rolled and stored. Thin coatings are durable than thick coatings since they have a tendency to crack.

Galvanised sheets are extensively used as roofing materials. Steel rods used as reinforcing rods in concrete are also galvanised to increase their resistance to corrosion. Similarly, mild steel pipes are also galvalised to increase their life.

Zinc is toxic and hence food containers are not coated with zinc. Further zinc is readily attached by acids so that galvanised sheets should not be exposed to acidic environment.

(b) Tinning (Cathodic coating): Another hot dipping process is tinning used for coating metals like steel, brass, bronze with tin. Tin is cathodic to iron and hence tinning is also referred to as cathodic coating. As in the case of galvanising, the metal surface is thoroughly cleaned and pickled in dilute H_2SO_4 solution, to remove the oxide layer.

After washing and drying the sheet, it is passed through a bath of molten tin which is covered with a layer of zinc chloride which acts as flux. The flux facilitates formation of thin coating of tin on the surface of the metal. The coated sheet then passes through bath of oil which prevents the oxidation of coated surface. The rollers squeeze out the excess metal and makes the coating uniform and adhering. Tin is non-toxic and hence tinning is extensively used for containers used for storing edible oils, ghee, fruit pulp, juices, kerosene, etc.

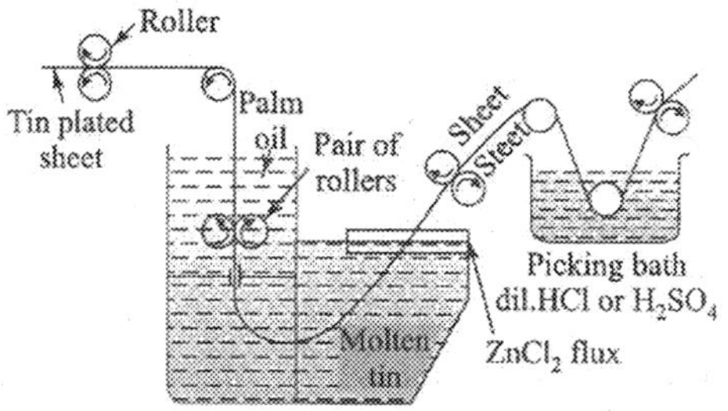

Fig. 1.14 Tinning.

1.7.6.2 Metal cladding

Metal cladding is method of coating the surface of base metal with another metal having more corrosion resistance. The process involves pressing together, the two metal sheets by passing them together through hot rollers. The metal sheets get bound together. Generally, the coating metals form just 5% of the composite cladded metal. The base metal can be cladded either on the single side or on both sides as required. In all cladding, the composite metal containing the cladding metal behaves as cladding metal especially with regard to corrosion resistance properties. Mild steel, aluminium, copper are some of the base metals which are cladded. Cladding metals are nickel, nickel alloys, silver, copper and lead alloys, stainless steel. Copper cladded aluminium cooking vessels have shown increased heating efficiency. Copper cladded steel is used in the electrical industry because of its good electrical conductivity and high strength.

The importance of cladding is due to cost difference in using only a thin surface on the base metal instead of thick section of costly metal, to achieve the same results. A thin sheet of duralumin is sandwiched between two thin sheets of aluminuium.

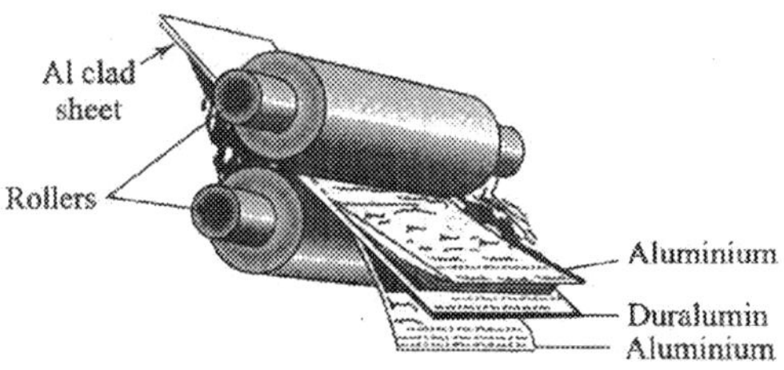

Fig. 1.15 The cladding of duralumin.

1.7.6.3 *Metal spraying*

The method is applicable for coating of low melting metals like Zn, Pb, Sn, Cu, Ni. The coating metal in the molten condition is sprayed on the metal surface with the help of a spray gun when a hard and wear resistant coating is formed. Large surface can be covered in short time and desired thickness of coating may be obtained easily.

1.7.6.4 *Diffusion coatings or cementation*

A protective alloy layer is formed on the surface of the base metal when it is heated in a drum packed with the powder of coating metal. The coating metal diffuses on to the surface of the base metal forming alloy layers of different compositions. The base metal is generally steel and coating metals used are Zn, Cr, Al.

- Coating of Zn is called *sherardizing*, the base metal articles like nuts, bolts, etc. are heated to 300-450°C in the presence of Zinc powder in an inert atmosphere in a closed drum which is rotated for 3 to 12 hours. The thickness of the coating can be controlled by duration of heating. Sherardizing does not alter the dimensions of the article.
- Coating of chromium is called as chromising. The article like gas turbine blades, etc. is packed in a drum containing 45% alumina and 55% Cr and heated in an inert atmosphere at 1300°C to 1400°C for 3 to 4 hours. Alumina prevents the coalescence of the chromium particles. Chromising improves resistance to salt sprays, steam, dilute nitric acid atmosphere.
- Coating of alumina is known as colourizing, the article is heated in a mixture of aluminium powder and alumina with little ammonium chloride

as a flux in an atmosphere of hydrogen for 4-6 hrs at 840° – 940°C. This treatment gives resistance to high temperature oxidation and sulphureous gases. It is done for furnace parts, condensers, retorts, valves etc.

1.7.6.5 Vapour depositions

The metal is vaporized by direct heat or by electric arc and is made to deposit on the surface of the base metal.

Application of high voltage (500 to 2000 volts) between anode and cathode causes disintegration of the cathode metal Ag, Cu and vapours are made to condense on the surface of the base metal. This method is known as cathode-sputtering. Decomposition of certain metallic carbonyls, halides etc. result in the deposition of the metal.

1.7.6.6 Electroplating or electrodeposition

Electroplating is a process in which metals are deposited or plated on base metals by passing an electric current through an electrolytic solution containing soluble salt of coating metal. Purpose of electroplating is to remove corrosion resistance and to obtain decorative articles.

1. Electroplating Process

The freshly cleaned article which is to be coated is made the cathode in a suitable electrolyte bath containing:

- a solution of the salt of the metal to be electrodeposited
- suitable pH is maintained.

The concentration of the salt solution is maintained by the addition of the metal salt at regular intervals or by the use of continuously dissolving anode of the metal. The plating is usually done at a high current density. The nature of the deposit depends upon the current density, pH and the concentration of the bath. In order to obtain a bright and lustrous deposit, specific chemicals are added to plating baths. The agents added are called brighteners. Aromatic sulphones, sulphonates, thiourea, coumarin, etc. are used for brightness in nickel plating.

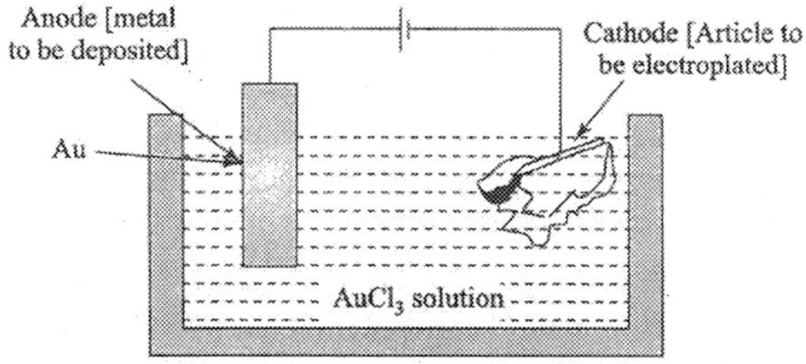

Fig. 1.16 Electroplating of gold.

Different types of electroplating are as follows:

(a) Chromium plating: For protecting iron, chromium plating is widely used. A warm bath of chromic acid and a high current density is used. The anode is lead and the cathode is the base metal to be plated. Chromium coats give an excellent shining appearance and are stable because of the formation of an invisible, strongly adherent protective layer of Cr_2O_3 on exposure.

(b) Nickel plating: A bath of nickel ammonium sulphate and aluminium anode is used. Nickel coats are not quite stable, the highly polished surface deteriorates with age with the appearance of rust spots.

(c) Copper plating: A bath of acidified copper sulphate solution and copper anode is used. The base metal is used as the cathode.

2. Factors Affecting Electroplating

- Cleaning of the article is essential in order to obtain good and uniform electroplating.
- Concentration of the electrolyte.
- Low concentration of metal ions will produce uniform coherent metal deposition.
- Thickness of the deposit should be minimised in order to get a strong adherent coating.
- Additives such as glue, boric acid, etc. should be added to the electrolytic bath to get a strong adherent and smooth coating.
- The electrolyte selected should be highly soluble and should not undergo any chemical reaction.
- pH of the electrolytic bath must be properly maintained to get the deposition effectively.

3. Applications of Electroplating

(i) Plating for decorative purposes: Silver, gold, copper, nickel, chromium, etc. are the most widely used metals for decorative plating. On the articles of steel, zinc, lead, copper, aluminuium alloys, etc.

(ii) Plating for protection: Against corrosion in order to protect steel from rusting and chemical attack, it is electroplated. Protective metals applied are zinc, cadmium, tin, etc. If both protection and good appearance are desired, steel is given successive coating layers of copper, nickel and chromium. Copper and nickel provide protection against corrosion. The chromium plate gives a clean metal appearance. In case of food containers, tin plating is preferred steel in order to avoid contaminate the food.

(iii) Plating for special surface and engineering effects: Electroplating has wide applications in engineering. A composite part can be designed in which the base metal is an alloy. It possesses the required mechanical properties but has poor resistance to surface attack under severe conditions. The required surface stability can be provided by means of electroplating.

Example 1: When a current of 0.5 ampere is passed through $CuSO_4$ solution, calculate time required to deposit 0.15 gm of copper. The Eq. wt. of copper is 31.6.

Solution:

$$W = Z \times C \times t$$

$$W = 0.15 \text{ gm}$$

$$C = 0.5 \text{ ampere}$$

$$t = 9$$

$$Z = \frac{\text{Eq. wt.}}{96500} = \frac{31.6}{96500}$$

$$t = \frac{W}{C \times Z} = \frac{0.15 \times 96500}{0.5 \times 31.6}$$

$$= 916 \text{ seconds.}$$

Example 2: A metal salt solution was electrolysed for 150 min by passing current 0.15 amperes. The weight of metal deposited was 0.783 gm. Calculate Eq. wt. and valency of the metal. (At. wt. of metal – 112)

Solution:

$$C = 0.15 \text{ ampere}$$

$$t = 150 \times 60 = 9000 \text{ seconds}$$

$$Q = Ct = 0.15 \times 9000 = 1350 \text{ coulombs}$$

∴ 1350 coulombs of electricity will deposit 0.783 gm metal

∴ 96500 coulombs of electricity will deposit $= \frac{0.783}{1350} \times 96500$

$$= 55.98 \text{ gm}$$

$$\text{Eq. wt. of metal} = 55.98 \text{ gm}$$

∴ $\text{Valency of metal} = \frac{\text{At. wt}}{\text{Eq. wt}} = \frac{112}{55.98} = 2.$

Example 3: Same quantity of electricity is passed through copper sulphate and silver nitrate solutions. If 0.29 gm copper and 0.99 gm of silver are deposited. If the Eq. wt. of copper is 31.6, find Eq. wt. of silver.

Solution: According to Farada'y second law of electrolysis

$$\text{Wt. of Ag deposited} = 0.99 \text{ gm}$$

$$\text{Wt. of Cu deposited} = 0.29 \text{ gm}$$

$$\text{Eq. wt. of Ag} = x$$

$$\text{Eq. wt. of Cu} = 31.6$$

$$\frac{\text{Wt. of Ag deposited}}{\text{Wt. of Cu deposited}} = \frac{\text{Eq. wt. of Ag}}{\text{Eq. wt. Cu}}$$

$$\frac{0.99}{0.29} = \frac{x}{31.6} = 107.875.$$

1.7.6.7 Chemical conversion coatings

Corrosion resistance of metals can be improved by developing stable surface films on metallic surfaces. These are formed by chemical or electrochemical reactions. These surface films can also serve as a good base for painting and giving other protective coatings.

The following are some important chemical conversion coatings:

Chemical Oxide Coatings

When the base metal is treated with alkaline oxidising solution, a thick layer of the oxide film is formed. For example, heating steel in air at 200°C to 450°C produces yellow or light blue colouring layer on the surface. Immersing steel in caustic alkali containing oxidising agents like $NaNO_3$, $NaClO_3$ produces black finish coating. Blue or green coloured coatings are obtained on copper by spraying hot ammonium sulphate solution on copper. Oxide coating serves as a base for painting.

Anodised Coatings

Anodised coatings are formed by oxidation process on metals. Such coatings are obtained on metals like Al, Zn, and Mg. A thick protective oxide film on Al is obtained by electrolysing a solution of sulphuric, chromic or phosphoric acid at a temperature of 35 to 45°C using Al as the anode at moderate current densities. The layer is formed by progressive oxidation on the surface of the metal. Anodised coatings on Zn, Mg can be formed similarly.

Phosphate Coatings

They are obtained by immersing the base metal in a solution of phosphoric acid. Such coatings can be obtained on iron, steel and zinc. These coatings do not give complete resistance to corrosion but serve as a good base for painting.

Vitreous Porcelain Enamel Coating

Vitreous or porcelain enamel are glass like protective and decorative film applied to steel and iron. Enamelled cast iron is widely used for bath tubs, washbasins, sanitary wares in hospitals. Enamelled steel is also used in refrigerators, kitchen, utensils, etc.

Chromate coating

Chromate coating can be obtained on metals like Zn, Al, Mg, etc. by immersing the base metal in acidified potassium chromate solution and then in a solution of neutral potassium chromate solution. Chromate films are of various colours, amorphous, non-porous and better in corrosion resistance than the phosphate films.

1.7.6.8 Organic coatings

Organic coatings are inert organic barriers on surface of base metal for corrosion resistance and decoration. They are composed of organic substances which form an inert, continuous and impervious layer on metallic surfaces. These coatings on account of their impermeability to water, salts and gases offer protection to the metal in the presence of the corrosive environment by acting as physical barrier.

Paints

A paint is a viscous suspension of finely divided pigment in a vehicle along with other ingredients like drier, thinner, and an extender. Paints form adhesive film which serve the purpose of protection as well as decoration.

Characteristics of a good paint

- Should be corrosion resistant
- Should have moderate viscosity for easy application
- The film should not crack on drying
- Should have good apperance
- If should adhere well to the surface
- If should form good and uniform film on surfaces.

Pigments

A pigment is a solid and colour producing substance in the paint.

Characteristics: A pigment should be chemically inert, non-toxic, opaque and possess high covering power, cheap and easily available and freely with film forming constituents of the oil.

The most commonly used pigments are:

- White pigments – white lead, $BaSO_4$ and TiO_2
- Blue pigments – prussian blue and ultramarine blue
- Black pigments – graphite and carbon black
- Red pigments – red lead and F_e2O_3
- Green pigments – chromium oxide
- Brown pigments – iron containing clays
- Yellow pigments – chrome yellow and zinc yellow

Functions of the pigment

- It gives opacity and colour to the film.
- It provides and aesthetic appeal to the film.
- It gives strength to the film.
- It protects the film by reflecting the destructive ultravoilet rays.

Vehicle or medium or drying oils

The liquid portion of the paint in which the pigment is dispersed is called a medium or vehicle. This is the film forming constituent of the paint. These are high molecular weight fatty acids present in animal and vegetable oils, e.g. linseed oil, dehydrated caster oil, soya bean oil, fish oil, dehydrated castor oil, etc.

Functions

- They hold the pigment particles together on the metal surface.
- They form the protective film by evaporation or by oxidation and polymerization of the unsaturated constituents of the oil.
- They give better adhesion to the metal surface.
- They impart water repellency, durability and toughness to the film.

Thinners

Thinners are added to the point to reduce the viscosity of the plants so that they can be easily applied to the metal surface. Thinners are volatile in nature and evaporate easily after application of the paint, e.g. turpentine, petroleum spirit, napth, benzene, zylol, etc.

Functions

- They reduce the viscosity of the paint to render it easy to handle and apply to the metal surface.
- They dissolve oil, pigments and produce a homogeneous mixture.
- They increase the elasticity of the film.

Driers

They are catalytic substances used to accelerate the drying of the oil film. They absorb oxygen from air and thereby act as catalysts. Co, Mn, Pb and Zn napthanates, borates and tungstates of Pb, Co, Mn are used.

Functions

They act as oxygen carrier catalysts which help the absorption of oxygen and catalyse the drying of the oil film.

Plasticizers

Addition of plasticizers improves the elasticity and flexibility of the film and also prevents the cracking of film. They are non-volatile, chemically inert and stable. Some of the substances used are triphenyl phosphate, diamyl phthalate, tricresy/ phosphate and high molecular weight esters.

Extenders or Fillers

Inert substances used to reduce the cost and to increase the durability and mechanical strength of the paint film. They prevent the settling of heavy pigments and also prevent the cracking of the paint films. Substances like $BaSO_4$, $BaSO_3$, $CaCO_3$, $CaSO_4$, clay, talc, mica, asbestos, magnesium silicate, etc. are used as fillers.

Antiskinning Agents

They prevent the skinning of the paint film, e.g. polyhydroxy phenols.

1.7.6.9 Organic coating varnishes

Varnish is a colloidal solution obtained by dissolving natural or synthetic resin in solvents or they are dispersions of synthetic or natural resin in oil and/or thinner.

They normally do not contain pigments. They may contain other constituents like drier or antiskining agents. They dry by evaporation of the solvent and by oxidation and polymerisation of part of the constituents present in the oil. There are essentially two types of varnishes. They are oil varnishes and spirit varnishes.

Oil Varnishes and Oleoresinous Varnishes

Oil varnishes and oleoresinous varnishes are prepared by dissolving resin in oil and volatile solvent. The oils used are linseed oil, tung oil, fish oil, soya bean oil, etc.

Resins used are natural like rosin, shellac, copal, etc. and synthetic resins like phenol formal dehyde, urea formaldehyde, alkyl resins, vinyl resins, etc.

Driers used are naphthenates, linoleates of cobalt, manganese, zinc and lead.

Thinners used are turpentine, naphthas, alcohol, kerosene etc.

The actual properties of oil varnishes are dependent on the types of drying oils and the types of resins used for the manufacture. They are used for exterior as well as interior work.

Spirit Varnishes

They are prepared by dissolving resins in a completely volatile solvent. They dry by evaporation of the solvent. On drying, the films developed are brittle and have a tendency to crack and peel off. To prevent this, plasticisers are added. The commonly used varnishes are prepared by dissolving the natural resin shellac in alcohol.

Uses

Varnishes are used on articles for maintaining a natural finish and for improving the natural appearance. They are also used as a brightening coat on painted surfaces and polishing wooden furnitures.

1.7.6.10 Enamels

They are pigmented varnishes and pigmented lacquers containing oil, drier and a thinner. They can be prepared in all colours. They have good covering power and possess the advantages of paint and varnish. On drying, they form lustrous and hard coatings. The pigments used are very fine textured and easily dispersible. They may be used with or without extenders.

In oleoresinous vehicles, the resins used are phenol, aldehyde resins, esters, etc. and the drying oils used are linseed oil, soya bean oil, fish oil, dehydrated castor oil, etc. Different resins give various degrees of hardness, washability and colour stability to the film. Driers used are naphthenates, resinates and linoleates of Pb, CO, Mn and Zn are used. Terpentine, xylol, acetone, etc. are used as thinners.

1.7.6.11 Lacquers

A lacquer is a colloidal dispersion of cellulose derivatives, a resin and a plasticizer in a volatile solvent. When applied on surface, a lacquer dries slowly forming a transparent, hard and waterproof coating. The addition of natural or synthetic resin increases the thickness and hardness. The addition of plasticizers decreases the brittleness and improves the flexibility and adhesion properties. The common plasticizers used are camphor, castor oil, dibutyl pthalate, tricresyl phosphate. The solvents used are ethyl acetate, butyl acetate, acetone, etc.

Lacquers are applied by spraying. They are used in furniture and automobile industries. They also find use in interior decoration and corrosion resistance.

QUESTIONS

- What is corrosion? Explain the term giving suitable examples.
- How do metals undergo corrosion? Explain the dry corrosion.
- What is electrochemical corrosion? Explain the mechanism of electrochemical corrosion.
- Explain the terms corrosion and erosion. Give a brief account of the electrochemical corrosion taking place in the case of metals of different electrode potentials.
- Define the term electroplating. Explain electroplating of gold.
- Mention and explain the various factors that influence the rate of corrosion of metals.
- How is the rate of corrosion influenced by the following factors:
  (a) Design of the material
  (b) pH
  (c) Purity of metal
- What is differential aeration? Give suitable examples to illustrate differential aeration corrosion.
- What is corrosion? Discuss the mechanism of wet corrosion.
- Discuss the different methods used for controlling the corrosion of metals.
- How can corrosion be controlled by the following methods:
  - Control of the environment
  - Cathodic protection
- Discuss the method of controlling corrosion by proper design and material selection.
- Mention the various methods used for prevention of corrosion and discuss the use of sacrificial anode for preventing corrosion.
- What is a paint? State ingredients of a paint with its functions.
- What is stress corrosion? How can it be prevented?
- Write notes on the following:
  (a) Anodic protection
  (b) Paints and varnishes

(c) Anodic and cathodic coatings
(d) Metal spraying and metal cladding
(c) Tinning and galvanising

- Explain the cathodic protection method of corrosion control.
- Write notes on (a) Corrosion control by proper design (b) differential aeration.
- What are the factors that control the increasing sensitivity of an alloy to intergranular corrosion?
- What is meant by cathodic protection? Explain the two types of cathodic protection and their applications.
- Describe how electrochemical properties are used in protecting submerged metals from being corroded.
- What are the factors influencing corrosion? Mention the methods to overcome the same.

MUMBAI UNIVERSITY QUESTIONS

May 2002

- Silver, gold and platinum do not undergo oxidation corrosion.
- Explain corrosion due to the formation of concentration cell.
- What is corrosion? Explain corrosion due to the combination of different metal electrodes.
- How corrosion can be controlled by hot dipping and metal spraying?

Dec. 2002

- Explain sacrificial anode method for protection of metal from corrosion.
- Write three differences between paint and varnish.
- Explain with equation the mechanism of wet corrosion.

May 2003

- What is corrosion? What are the factors affecting the rate of corrosion?
- How does design and selection of material help in protection against corrosion?
- Explain the electrochemical theory of corrosion.
- Write a short note on a galvanic corrosion.
- Explain the cathodic method of protection against corrosion giving suitable diagrams.

Dec. 2003

- Why are tinned containers preferred to galvanised ones for storage of food products?
- What is corrosion? What are the factors that affect rate of corrosion?
- Mention various methods used for the application of metallic coatings. Explain galvanising and tinning in detail.

May 2004

- Differentiate between galvanising and tinning.
- How the proper design will reduce the rate of corrosion?
- Write a note on lacquers.
- Write a note on differential aeration corrosion.
- Explain how are the following factors responsible for corrosion:
  (i) Relative areas of anode and cathode (ii) pH (iii) humidity
- Differentiate between cathodic protection and anodic protection.

Dec. 2004

- Differentiate between galvanizing and tinning (any two).
- Describe the basic principle of electrochemical theory of corrosion. Explain rusting of iron with the help of this theory.
- State major constituents present in a paint. Mention their function.
- Write a short note on electroplating
- Write a short note on the following:
  (1) Sacrificial anodic protection (ii) Passivity of metal

May 2005

- Give the characteristics of a good quality paint.
- What is metal cladding?
- Why galvanisation of articles is preferred to tinning?
- How are the following factors responsible for rate of corrosion:
  (i) Relative areas of anode and cathode (ii) pH (iii) overvoltage (iv) nature of the surface film
- What do you mean by impressed current cathodic protection?

Dec. 2005

- Distinguish between anodic and cathodic metallic coating.
- What is corrosion? Explain the corrosion due to combination of different metal electrodes.

- Write a short note on cathodic protection by impressed current.
- What are the main constituents of a paint? Write their functions.

May 2006

- Explain how would you control corrosion by improving the design.
- How do the following influence the corrosion rate:
  (i) Position of metal in galvanic series
  (ii) Relative areas of anode and cathode
  (iii) Temperature
  (iv) pH of the medium
- Mention the various methods of metallic coating and explain any two of them.

Dec. 2006

- Give reason – steel pipe connected to copper plumbing corrodes.
- Explain how corrosion can be prevented by the following methods: (i) Cathodic protection (ii) Design and material selection
- Distinguish between anode and cathodic coating.
- Discuss the electrochemical theory of corrosion.

May 2007

- Distinguish between galvanising and tinning.
- Write a short note on cathodic protection.
- Explain the following factors influencing the rate of corrosion:
  (i) Relative area of cathode and anode (ii) pH of medium
- What are the main constituents of a paint? Write their functions.

Dec. 2007

- Explain with examples, how the nature of the oxide film formed on the surface of metal influences the rate of corrosion.
- Explain how corrosion can be prevented by proper design and selection of material.
- Explain rusting of iron with the help of electrochemical theory of corrosion.

May 2008

- Define corrosion. List various types of corrosion. (3)

- Distinguish between galvanising and tinning. (4)
- How are following factors responsible for the rate of corrosion:

  (i) Relative areas of anode and cathode (ii) pH of medium (iii) Nature of surface film. (6)
- Define paints? State the characteristics of a good paint. (5)
- A current of 2 Amp. was passed through a solution of copper sulphate for 16 minutes. Calculate amount of copper deposited on cathode? (At. wt. Cu = 6z3) (5)

CHAPTER 2

ALLOYS

2.1 INTRODUCTION

An alloy is an intimately mixed mixture of two or more elements out of which one is essentially a metal. They are classified into:

(a) Ferrous alloys: These contain iron as a major component, e.g. steels.

(b) Non-ferrous alloys: These alloys do not contain iron as a major component, e.g. Al, Cu, Pb, Sn alloys.

2.2 NON-FERROUS ALLOYS

These alloys contain metals like copper, aluminium, zinc, nickel, tin and lead as the principal alloying constituent. The characteristics/properties are:

- They possess attractive colours, softness and lower melting points as compared to ferrous alloys.
- They have good corrosion resistance and low density.
- They have high electrical and thermal conductivity.
- They are easily castable and can be cold worked.
- They have low coefficient of friction and can be fabricated easily.
- Most of the alloys have wide applications in the engineering fields.

2.2.1 Alloys of Copper

These alloys contain copper as the principal constituent e.g. brasses and bronzes.

2.2.2 Essential Properties of Copper

The copper metal is highly resistant to corrosion. It is ductile, malleable, having moderate to high hardness and strength. It possesses good machinability and can be soldered, brazed or welded. It has high thermal and electrical conductivity and can be easily polished or plated. Copper can be alloyed with elements like zinc, tin aluminium, lead, silicon, nickel, etc. and the alloys can be cast to obtain various shapes. These are low and high alloys of copper.

The low alloys of copper contain certain elements in minor quantities. High alloys of copper contain substantially larger percentages of alloying elements. Brasses and bronzes constitute the important alloys of copper.

2.2.3 Types of Brasses and Bronzes

In general, brasses contain 60 to 90% copper and 10 to 40% zinc. They have golden yellow colour and greater strength than copper. They have lower melting points than those of copper and zinc. They are fairly corrosion resistant in presence of water.

They are very commonly used for condenser tubes, sheets, cartridges and also utensils.

2.2.4 Some of the Main Varieties of Brasses

1. Commercial Brass

Composition: 90% Cu, 10% Zn

Properties

Golden yellow colour, stronger and harder than copper.

Uses

It is use for making hardware, screws, rivets, jewellery, etc.

2. Dutch Metal or German Brass (80/20 brass)

Composition: 80% Cu and 20% Zn

Properties

(a) It has a brilliant golden colour.

(b) It can be rolled into sheets or hammered into thin foils.

Uses

(a) It is used for making artificial jewellery.

(b) It can also used in musical instruments, battery cases, etc.

3. Muntz Metal (60/40 brass)

Composition: 60% Cu and 40% Zn

Properties

(a) It is suitable for only hot working.

(b) It is stronger than cartridge brass.

Uses

(a) It is used for making marine fittings, condenser tubes, valves.

(b) It is also used for making springs and chains.

4. Cartridge Brass

Composition: 67% Cu and 33% Zn.

Properties

It has good ductility and strength and can be cold-worked.

Uses

(a) It is used for making cartridge cases, condenser tubes.

(b) It is used in household articles.

2.2.5 Special Brasses

These brasses contain certain other metals in addition to copper and zinc.

1. Admiralty Brass

Composition: 71% Cu and 28% Zn and1% Sn.

Properties

(a) It possesses good strength, hardness, corrosion resistance due to addition of tin.

(b) It is corrosion resistant to sea-water.

Uses

(a) It is used for marine work.

(b) It is used for making motor boats and pump parts.

2. Naval Brass

Composition: 62% Cu, 37% Zn and 1% Sn.

Properties

It has good abrasion and corrosion resistance.

Uses

(a) It is used in marine works.

(b) It is used for making condenser tubes.

3. German Silver Composition

60% Cu, 15% Zn and 25% Ni.

Properties

(a) It has attractive white colour.

(b) It has good resistance to corrosion.

Uses

(a) It is used for making decorative articles, tablewares, etc.

(b) It is also used making coins and utensils.

4. Lead Yellow Brass

Composition: 1% Sn, 1-3% Pb, 24 to 36% Zn and the remaining copper. It may also contain 0.3% Al.

Uses

It is used for hardware, valves, ornamental castings, etc.

5. Silicon Brass

Composition: 14% Zn, 3.5 to 4% Si, 0.5% Pb and the remaining part copper.

Uses

It is used in die cast parts, gears and structural parts.

2.2.6 Important Bronzes

1. Gun Metal

Composition: 5-10% Sn, 2-5% Zn, and the remaining copper.

Properties

It possesses good tensile strength and resistance to corrosion.

Uses

It is used for bearings, steam pipe fittings, marine castings, hydraulic valves, gears, heavy load bearings, etc.

2. Phosphor Bronze

Composition: 93% to 96% copper, 4-7% Sn and P less than 0.4%.

Properties

(a) It is hard, tough and possesses good strength.

(b) It possesses good casting properties and has a low coefficient of friction.

(c) It shows good corrosion resistance particularly to sea water.

Uses

(a) Bronzes with low phosphorus content are used in making springs, electric switches, radio aerial wires and suspension wires for moving coil galvanometers, spindles for valves and pumps, steam fittings in boilers, etc.

(b) Bronzes containing higher percentages of phosphorus and tin (10%) are suitable for purposes where stresses occur. Due to its increased hardness and abrasion resistance, they find use in bearing materials and gears. They are also used for turbine blades, taps, bushes, etc.

3. Aluminium Bronze

Composition: 89-93% Cu and 9-10% Al and minor quantities of elements like Fe, Mn, Ni, etc.

Properties

(a) It has a good golden yellow colour.

(b) It is quite tough and gives good castings.

(c) It possesses good abrasion resistance.

(d) It is resistant to atmospheric corrosion and also corrosion at high temperature.

(e) They possess good tensile strength and shock resistance.

Uses

(a) They are useful for all casting operations, bearings and bushes.

(b) They are useful for making coins, utensils, artificial jewellery and photo frames.

4. *Nickel Bronze*

Composition: 90% Cu, 9% Ni and 1% Fe.

Properties

(a) They are hard and have better tensile strength.

(b) They have better corrosion resistance compared to that of copper.

Uses

(a) It is used for general purposes.

(b) It is used for making hard bearings and unhardened shafts and valves.

5. *High Nickel Bronzes*

Composition: 15%–60% Ni, 12% Zn, upto 3% Sn, 0.1% Mg and remaining part copper.

Properties

(a) They are very hard and have very good tensile strength.

(b) They have high corrosion resisting properties.

Uses

They are used for hydraulic components used in contact with salty sea water.

6. *Bearing Bronze*

Composition: 80% Cu, 10% Sn, 10% Pb.

Properties

It is reasonably hard and possesses good corrosion resistance.

Uses

It is used in making bearings which resist heavy pressures as in railways.

7. *Bell Metal*

Composition: 80% Cu and 20% Sn.

Properties

It is hard and produces sonorous sound.

Uses

It is used in making bells and gongs.

8. Beryllium Copper

Composition: 98% Cu and 2% Be.

Properties

(a) It has good resistance to wear and corrosion.

(b) It has good conductivity.

(c) It is used in making brushes, relays and switches.

2.2.7 Alloys of Aluminium

Aluminium is a metal known for its light weight, high corrosion and non-toxicity. Its specific gravity is 2.7 with a melting point 658°C and boiling point 2450°C. The properties of aluminium are further improved by alloying with elements like Mg, Cu, Ni, Zn, Si, Mn, etc. The resultant alloys are also light and have good resistance to corrosion. They have good mechanical properties and high electrical conductivity. On account of these properties, they find extensive use in aircraft industry.

Two types of aluminium alloys:

Cast Alloys

These alloys contain copper and silicon as the important alloying elements. They may also contain Mg, Zn, Ni and Fe to a small extent. They are shaped by casting

Wrought Alloys

They contain Cu, Mn, Mg, Si in different amounts. They are non-heat treatable or heat treatable.

The non-heat treatable alloys contain Mg and Mn as major alloying elements. Their strength increases by cold working. They have good corrosion resistance and ductility.

The heat-treatable aluminium alloys improve in their strength by cold working as well as by heat treatment process. They can be heated to the required temperature and then quenched in oil or water for a required period of time to improve their strength and hardness.

The light weight aluminium alloys are characterised by malleability, ductility, high thermal-conductivity, non-magnetic properties, higher coefficient of expansion compared to steel, tensile strength equivalent to structural steel, and above all a good resistance to corrosion.

2.2.7.1. *Important alloys of aluminium*

1. Duralumin

Composition: 95% Al, 4% Cu, 0.5% Mn and 0.5% Mg.

Properties

(a) It is a light, ductile and tough alloy.

(b) It has good conductivity of heat and electricity.

(c) Though its strength is comparable to that of steel, its density is only one-third of steel.

(d) It possesses good machinability.

(e) Its tensile strength can be increased to 2000 kg/cm^2 by heat treatment without affecting its ductility.

(f) Its corrosion resistance can be improved by the process of cladding (one thin sheet of the alloy is sandwiched between the two thin sheets of pure aluminium).

Uses

(a) Due to its lightness and strength it is used in automobile and aircraft industries in the form of 'Alclad' sheets. It is also used in locomotive parts.

(b) Due to its good electrical conductivity and high ductility, it is used in making surgical instruments, cables, fluoresent tube caps, structural framework, hardware, body of vehicles, rivets, bars, etc.

2. Magnalium

Composition: 90-95% Al and 5-10% Mg.

Properties

(a) It is lighter than pure aluminium and magnesium.

(b) It is quite strong and tough.

(c) It is comparable to brass in its mechanical properties.

(d) It has good corrosion resistance.

Uses

It is used in making chemical balance, scientific instruments, furniture, aeroplane parts, etc.

3. γ-alloy

Composition: 91% Al, 4% Cu, 1.5% Mg, 2% Ni, 0.5% Fe, 0.5% Mn, 0.5% Si.

Properties

(a) It possesses high strength and hardness at high temperatures.

(b) It is an excellent light weight alloy, having very good corrosion resistance.

Uses

(a) It is used in aeroplane parts.

(b) It is used in heat exchange parts, refrigeration, storage containers, etc.

(c) It is also used in making cylinder heads and crank cases of internal combustion engines, overhead cables, etc.

2.2.8 Lead and Tin Alloys

2.2.8.1 *Solders*

Solders are lead-tin alloys having low melting points, used primarily for soldering purposes.

There are two types of solders: (a) Soft solders (b) Hard solders.

When an alloy with a lower melting point is used for joining, the process is known as soldering and when an alloy with higher melting point is used, it is called brazing.

Some of the important solder alloys are given below :

1. Soft Sloiders

Composition

Pb-67-37%, Sn-66-31%, Sb–0.12%.

Properties: Their melting points are in the range of 200-300°C.

Uses: They are useful for soldering electrical connections, joining lead pipes, sealing tin cans.

2. Silver – Solders

Composition

Ag-10-80%, Cu-15-42%, Zn-4-48%

Uses: They are used for brazing.

3. Tinman's Solder

Composition

Sn-66%, Pb-34%

Uses: Its melting point is 183°C and is used for soldering articles of tin.

Type Metal

Composition

Pb-75%, Sb-20%, Sn-5%

Properties: It expands slightly on solidification and gives good castings.

Uses: Mainly used for printer's type.

Woods Metal

Composition

Bi-50%, Pb-25%, Sn-12.5%, Cd-12.5%

Properties: M.P. – 70°C

Uses: It is used for making automobile sprinklers, safety plugs for cookers, milk pots, boiler and electric fuses, as castings in dental work.

Rose Metal

Composition

Bi-50%, Pb-28%, Sn-22%

Properties: It has a low point of 89°C and is easily fusible.

Uses: It is used for fire alarms, fuse wires, automobile sprinkler systems.

2.3 POWDER METALLURGY

2.3.1 Introduction

Metals and alloys can be fabricated in order to obtain desired shape by hot working or cold working processes. A new technique called powder metallurgy has been developed for fabricating metallic parts from metal powder. Powder metallurgy can be defined as the technique of producing metal powders, and fabricating them into desired shape of articles. This technique is used to make objects of high melting alloy metals, refractory composites, etc. Metal powders find use in various industries such as catalyst, paints, explosive, printing inks, etc.

2.3.2 Principle of Powder Metallurgy

Powder metallurgy is essentially the art of producing metallic powders of a single metal or different metals. The powders are thoroughly mixed, compacted at high pressure into a particular shape and then heated at elevated temperature below the melting point of the solid powder. The shaping by pressure is called compaction and heat treatment process is called sintering.

2.4 CHARACTERISTICS OF METAL POWDERS

For the manufacture of products by powder metallurgy, accurate knowledge of metal powders and their characteristics are needed. The important characteristics of the metal powders are as follows:

(1) *Purity*: The impurities present in a metal powder cause adverse effects on the products. Better mechanical properties can be obtained by using pure metal powders.

(2) *Density*: Density is mass per unit volume. Apparent density is the weight of a unit volume of a powder when packed loosely.

(3) *Particle size & shape*: The particle size used in powder metallurgy varies form 5 – 200 microns. The particle size influences the strength, density, and porosity of the product.

(4) *Compression Ratio*: It is the ratio of the volume of the uncompressed powder to the product after pressing. It is also defined as the ratio of the density of the compact to the apparent density of the powder. This ratio is important in designing of particular powders.

- Metal powders have various shapes like spherical, angular rounded etc. Spherical particles have excellent sintering properties whereas irregular-shaped particles have superior moulding properties.

(5) *Flow Factor*: It is defined as the rate of flow of the metal powder through a definite orifice. Good flow factor is necessary for filling the moulds quickly and uniformly.

(6) *Ability to be Sintered*: Sintering ability provides and promotes good bounding between all the particles in the product which results in better mechanical and physical properties of the product.

2.5 TECHNIQUES FOR PRODUCING METAL POWDERS

The various mechanical, chemical and electrochemical techniques of producing metallic powders of carbides, borides, nitrides, ceramics are described in the following sections.

2.5.1 Mechanical Process (Mechanical pulverisation)

Mechanical process can be used to prepare metal powders. They use counter rotating plates or rapidly moving hammers by which the mechanical forces disintegrate the particles to maximum fineness. Pulverisation is ususlly followed by ball milling. Other methods include machining and shotting. Machining gives coarse particles which may be further converted into fine particles by milling. Magnesium powder can be prepared by this method. In shotting, the molten metal is passed through a sieve or an orifice followed by cooling in water by dropping from a height. Spherical-shaped particles of aluminium, lead and zinc are produced by this method.

2.5.2 Atomisation

Atomisation is commonly used for metals which have low melting point-like Pb, Zn, Cd, Sn, etc. In this process, the molten metal is forced through a small orifice and is broken by a powerful jet of compressed air. As a result, the jet of liquid metal is changed into very small particles that are spherical in shape. The collision between the stream of metal and air produces very small particles which are called atoms and hence the process is called atomisation. This process requires careful control of temperature and pressure. Oxidation can be prevented by providing an inert atmosphere.

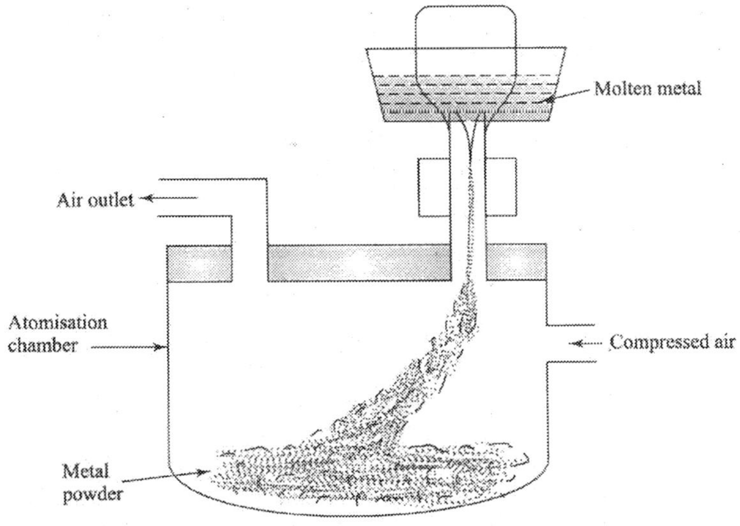

Fig. 2.1 Atomization.

Advantages of atomisation

- The metal powder obtained has high purity.
- Low initial cost and rate of production is high.

2.5.3 Electrolytic Deposition

Copper, silver and molybdenum powders are manufactured by this process. This process is similar to electrolysis. For making copper powder, copper plate is placed as anode and aluminium plate is placed as cathode in a tank of electrolyte. On passing electric current, copper powder deposits on the aluminium cathode and is scrabbed off. It is washed with water to remove the traces of electrolyte and then dried.

Advantages of electrolytic deposition

- Pure powder can be obtained by this process.
- It is an easily controllable process.

The only disadvantage of this process is its operating cost. It is very high and production rate is very low.

2.5.4 Chemical Methods

Reduction, precipitation and condensation chemical methods are used to produce metallic powders.

Reduction: High melting metallic powders of metals like copper, cobalt, nickel, tungsten, molybdenum, etc. are produced by this method. In this process, metallic oxides are heated in a current of hydrogen to reduce them to metallic powders (Fig. 2.2). The spongy powder obtained is suitable for cold pressing due to its softness and plasticity.

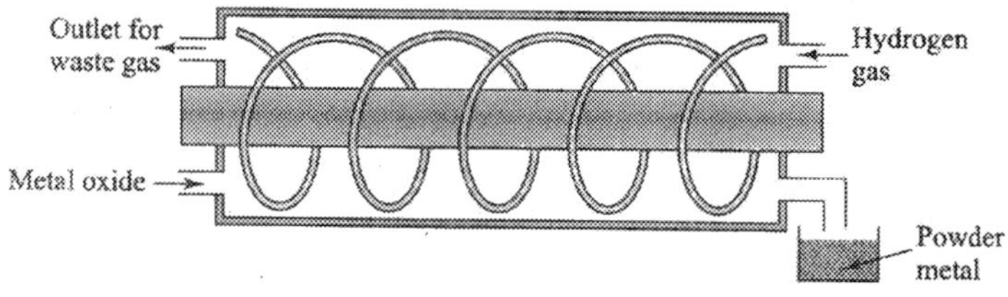

Fig. 2.2 Reduction process.

Precipitation: Extremely pure metal powders of silver, nickel and iron are produced by this method. Iron and nickel react with carbon monoxide to produce carbonyl $Fe(CO)_5$ and $Ni(CO)_5$ under suitable temperature and pressure. This liquid carbonyl decomposes resulting in metal precipitation at reduced pressure and elevated temperature. The precipitate is then dried and condensed to form powder.

Condensation: This method is used to produce zinc, cadmium and magnesium powders. A metal rod is heated on a high temperature flame where it gets vaporised. The vapours of the metal are allowed to condense by passing through a cold surface of the material. The pure metal condenses in the form of fine powder. It is not economical for large-scale production of powder.

2.6 VARIOUS STEPS INVOLVED IN POWDER METALLURGY

Production of Metallic Powders: Metal powders are produced by mechanical pulverisation, atomisation, electrolytic deposition or chemical methods.

Mixing or blending of powders: Uniformity in the finished product can be achieved by proper mixing of powders. To reduce die wall friction lubricants are added in most of the powders during mixing e.g., stearic acid, lithium stearate, etc.

Pressing or Compacting: Metal powder after mixing is pressed to shape in steel dies under pressure at room temperature. The solid mass obtained is known as green compact. This process of converting loose powder into green compact of accurate size and shape is known as compacting.

The compaction methods are subdivided into cold compaction and hot compaction.

(a) Cold compaction method: It takes place in the ambient range of temperature within which high temperature deformations like diffusion or dislocations are negligible. Several methods of cold compaction are available.

(i) *Cold pressing:* It is the most important compaction method in powder metallurgy. The starting material is the bulk powder with or without small amount of binder or lubricant. The cold pressing process may involve axial die pressing or isostatic pressing principles.

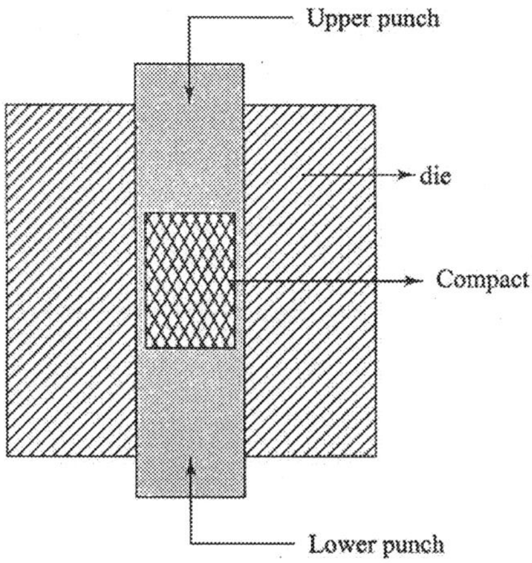

Fig. 2.3 Axial pressing.

In axial pressing, the powder is compacted in rigid dies by axially loaded punches. The main process variable in this technique is the axial compaction pressure which is defined as the ratio of the punch load and the punch surface area.

(ii) *Powder injection moulding:* It is well known process in the large scale manufacture of parts from thermoplastic polymers. When this process is applied for powder metallurgy it is called metal injection moulding. Steps involved in PIM.

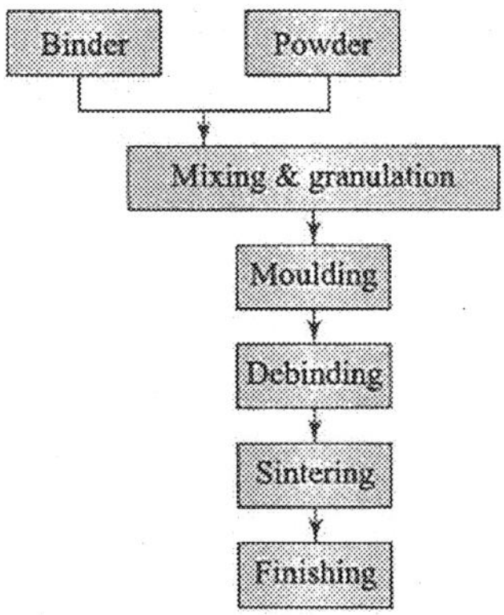

(iii) *Cold-powder Extrusion:* This is plastic forming method (similar to injection moulding) in which a plasticised powder system is forced by a piston or screw unit into forming die, by using typical equipment. Tubes and rods can be prepared by this method.

(iv) *Roll compaction:* In this method, metal powders are compacted in the gap between two rollers. This is an alternative technique for producing semifinished products like strips and sheets. Production of high purity strips and sheets of cobalt and nickel is most notable commercial application of this process. The strips are used for making coin blanks for the mint, electronic tubes, batteries, thermostats and other bimetal applications.

(b) Hot compaction: In this process, the deformation mechanism of the powder are activated by simultaneous applications of higher processing temperature and pressure. The major hot compaction techniques:

(i) Axial hot pressing
(ii) Hot forging
(iii) Hot extrusion
(iv) Isostatic hot pressing.

Effects of compacting are:

(a) The porosity of the material decreases and the surface irregularities are removed as the particles are forced together under pressure.
(b) Compaction increases tensile as well as the compressive strength of the material.
(c) Cold welding occurs between adjacent particles.

Presintering: This process involves heating of the green compact below the sintering temperature to remove the lubricants added during blending.

Sintering: The mechanical and physical properties can be improved by heating the compacts in a furnace under controlled atmospheric conditions. The particles are pressed into a more compact mass by the application of heat. During sintering the fluidity of the component increases. This results in better interlocking of atoms of the constituents. Gases and other volatile materials present in the constituents are also eliminated by heat. The temperature of sintering is always kept below the melting point of the metal. The temperature and time of sintering of some materials are as follows.

Type of powder	*Temperature in °C*	*Time*
Aluminium and its alloys	400	3 hrs
Iron and its alloys	1100	2 – 3 hrs
Brass	900	5 hrs

Ceramic Powders: Naturally occurring silicate ceramics raw materials such as clay, kaolin are commonly used. The advanced ceramics requires pure and well-defined powders which are prepared synthetically. Manufacture of advanced ceramics is similar to powder metallurgy. Ceramic powders are synthesised by the following methods.

(1) Melting—Ceramic oxides like Al_2O_3 and ZrO_2 are manufactured by melting. This process is mainly used for producing coarse grade powders used in the manufacture of refractory materials.

- Manufacture of alumina: High purity aluminas are produced by the decomposition of high purity sulphates, chlorides, nitrides of aluminium. In Bayers process bauxite is used to produce powders of lesser purity. After separating Al_2O_3 from Fe_2O_3 and other oxides and by converting into sodium aluminate the latter is converted into $Al(OH)_3$ by hydrolysis which is then calcined to Al_2O_3. Applications in bioceramics, chinaware, electronic parts, special glasses, sodium vapour lamps, etc.
- Silicon carbide: The SiC was first prepared in 1981 by Acheson process by heating sand (60%) and coke (40%) at 1500°C in an electric furnace. 1-3% NaCl is used for purification.

$$SiO_2 + 3C \longrightarrow SiC + 2CO$$

SiC exist as hexagonal α-Sic and cubic β-SiC.

- Manufacture of ZrO_2 (Zirconia): It is manufactured from crude ZrO_2 or Zircon sands (Zr SiO_4) at 1370°K. It is refractory material with a melting point of 2950°K.

$$Zr\,SiO_4 + 2Na_2CO_3 \xrightarrow{1370^\circ K} Na_2ZrO_3 + 2CO_2$$

$$Zr\,SiO_4 + 4NaOH \xrightarrow{870^\circ K} Na_2\,ZrO_3 + Na_2SiO_3 + 2H_2O$$

- Reaction from solution:

$$Na_2ZrO_3 \xrightarrow{HCl} ZrOCl_2 \xrightarrow{hydrolysis} Zr(OH)_4 \xrightarrow{calcination} ZrO_2$$

These are very costly. It can withstand a temperature upto 2600°C. It is used in the manufacture of powder oxide ceramics of higher strength.

- Cermets: Hot pressed or sintered materials containing ceramic and metallic components.

Properties: High thermal conductivity and shock resistance used in the manufacture of ZrC, MoC, WC. They have excellent high temperature strength and hardness.

Uses: Used in rocket and jet engine parts as spinning tools, in the manufacture of powder oxide ceramics of higher strength etc.

Properties: High mechanical strength, dark grey coloured material having good hardness, fusion temp. 2500°C. Coefficient of thermal expansion is low. Specific gravity 3.17 to 3.2. In presence of air at 100°C it gives SiO_2 and CO.

$$2SiC + 3O_2 \longrightarrow 2SiO_2 + 2CO$$

Uses: Due to high thermal and electrical conductivity used in muffle and electric arc furnaces. It is used in the manufacture of powder oxide ceramics of higher strength (carborandom).

(2) Solid state reactions—Most of the oxide ceramics are produced by the thermal decomposition of hydroxides, oxalates, sulphates, carbonates, etc.

(3) Solid gas reactions—Oxides, carbides and nitrides are synthesized by the reaction of metals with O_2, hydrocarbons, N_2 or NH_3 respectively. For example,

$$3Si + 2N_2 \longrightarrow \underset{\text{Silicon nitride}}{Si_3N_4}$$

(4) Gas phase reactions—Ceramic powder having high specific surface area are produced by this method. This process consists of vapour phase decomposition or hydrolysis in a flame and specially used to produce SiO_2 and TiO_2 from $Ti-Cl_4$ and $Si-Cl_4$ respectively.

$$SiCl_4 + H_2O \longrightarrow SiO_2 + 4HCl$$

The most important non-oxide ceramics which are widely used are silicon carbide (SiC), silicon nitrides (Si_3N_4) and boron carbide (B_4C) are also synthesized.

2.7 APPLICATIONS OF POWDER METALLURGY

Lamp Industry: Tungsten metal has very high melting point and is very difficult to cast. Dr W.D. Coolidge developed tungsten powder by reduction of tungsten oxide with hydrogen. The tungsten filament is made from pure tungsten powder which is compacted and sintered to form a bar of tungsten. This tungsten bar is then converted into 2 mm or less in diameter and drawn through tungsten carbide dies to about 0.2 mm diameter which can be further reduced through diamond dies at about 650°C. This tungsten wire has more strength than the hardest steel. It is used as a filament in incandescent lamp industry.

Refractory Metal Composites (Cermets): The high melting metal powders like tungsten, molybdenum, etc. can be mixed with ceramic oxides to form refractory composites using powder metallurgy technique.

Diamond Tools: Diamond tools are prepared by blending 30 per cent diamond dust with iron powder. They are used for cutting hard materials like porcelain and glass.

Cemented Carbides: Tungsten, molybdenum, carbides are used to make cutting tools by powder metallurgy.

Porous Metal Sheets: Metal powders like copper, brass, bronze and stainless steel are rolled into porous sheets having controlled porosity using this technique.

2.7.1 Advantages of Powder Metallurgy

- This process makes cent per cent use of the material as there is no wastage of any raw material in casting.
- Any typical shaped article can be manufactured by this method.
- Relatively more uniform structure is obtained at high production rate.
- The articles made have good finished surfaces.
- The use of diamond in industry has been made possible mainly through powder metallurgy.

2.7.2 Disadvantages of Powder Metallurgy

- Many complicated shapes that can be produced by casting cannot be produced by powder metallurgy or limited applications.
- The cost of equipment required is very high.
- The physical properties of the material obtained by this process are lower than those obtained by other processes.
- Because of greater volume of metal powders they are difficult to store and get oxidized on long storing.
- The articles produced by this method have poor ductility.
- For small-scale production, this method is not economical.

QUESTIONS

- Define powder metallurgy. What are the various steps involved?
- Give reasons for tungsten filament prepared by powder metallurgy technique.
- Give applications of powder metallurgy.
- Write down the main steps for powder metallurgy process.
- What are the advantages and disadvantages of powder metallurgy process?
- Write a note on:

  (i) Atomization (ii) Compacting and sintering ceramic materials.
- Discuss the important copper alloys stating their composition, properties and uses.
- Write a note on alloys of aluminium giving their composition, properties and uses.
- What are brasses? Give their brief account of the varieties brasses giving their composition, properties and uses.
- Write an informative note on lead-tin alloys.

- Discuss four important types of bronze stating their composition properties and uses.
- Give composition, properties and uses of the following alloys:
  (i) German silver (ii) Duralium (ii)Tinmann's solder (iv) Magnelium (v) Rose metal (vi) Wood's metal (vii) γ-alloy (viii) Dutch metal (ix) Cartridge brass. (x) Commercial brass.
- What is powder metallurgy? What are the salient features?
- What are the advantages and disadvantages of powder metallurgy?
- How metal powders are prepared?
- Explain the processes involved in powder metallurgy.
- What are the applications of powder metallurgy?
- Explain the terms (i) Atomization (ii) Sintering.

MUMBAI UNIVERSITY QUESTIONS

May 2002

- Give composition, properties and uses of:
  (i) German silver (ii) Commercial brass.

Dec. 2002

- What are the various purpose of making alloys?
- Give the composition and uses of soft solders.
- What are various alloys of aluminium? Explain composition and uses of any two of them.

May 2003

- Distinguish between brasses and bronzes.
- What are the special effects of alloying elements on alloy steels?
- Give the composition, properties and uses of:
  (i) Nichrome, (ii) German silver

Dec. 2003

- Name two alloys of aluminium. State their composition and uses.
- State the effect of chromium, cobalt and tungsten on properties of alloy steels.
- What are alloys? Explain the purpose of making alloys.

May 2004

- Explain the effect of Co and Si on alloying of steel.
- Explain: Solders.
- Give the composition, properties and uses of:
  (i) Duralumin
  (ii) Nichrome.

Dec. 2004

- Write composition and application of German Silver.
- What are alloys? What is the purpose of making alloys?
- Write the composition and applications of the following:
  (i) Duralumin (ii) Hard Solders (iii) Monel metal

May 2005

- Give composition and properties of duralumin.
- What is the effect of following elements on alloying steel:
  (i) Cobalt (ii) Tungsten (iii) Nickel.
- Give an account of alloys of lead and tin.

Dec. 2005

- Give composition and uses of duralumin.
- Write the composition, properties and uses of the following:
  (i) German Silver (ii) Type metal (iii) Commercial brass

May 2006

- Give the composition, properties and uses of duralumin.
- Write composition, properties and uses of: (a) German silver (b) Nichrome.

Dec. 2006

- What are alloys? Explain giving at least two examples each, the purpose of alloying.
- Give the composition, properties and uses of duralumin.

May 2007

- What are alloys? What are the purposes of making alloy?

- Write composition, properties and uses of the following:
  (i) German Silver (ii) Duralumin

Dec. 2007

- What is the purpose of making alloys? Explain with examples.
- Write a note on solders.

May 2008

- Write composition, properties and uses of duralumin. (3)
- What is powder metallurgy? How are metal powders prepared? (6)
- Write composition and uses of Wood's method. (2)
- Write note on atomisation and sintering. (5)

CHAPTER 3

FUELS

3.1 INTRODUCTION

A fuel is a combustible substance containing carbon which is an important source of heat energy and power. Fuel on proper burning gives large amount of heat that can be used economically for domestic and industrial purposes. Substances like coal, oil and natural gas are available in different countries. Carbon compounds have been used for many centuries as the source of heat and energy.

The main sources of fuel are coal and petroleum. These are stored fuels available in the earth's crust and are generally called fossil fuels because they were formed from the fossilised remains of plants and animals. Apart from these, there are also nuclear fuels which generate energy by fission or fusion of nuclei.

Classification of fuels: Fuels are classified as

- Primary fuels which occur in nature as such, e.g. petroleum, natural gas, wood, peat, lignite, bituminous and anthracite coals.
- Secondary fuels which are derived from the primary fuels, e.g. coke gasoline, coal gas, charcoal, etc.

Both primary and secondary fuels may be further classified based upon their physical state as: (i) solid fuels (ii) liquid fuels, and (iii) gaseous fuels.

Table 3.1 Fuels

Physical state	*Natural or primary*	*Prepared or secondary*
Solid	wood, peat, lignite	coke, charcoal
Liquid	petroleum oil	petrol, kerosene, diesel, molten oil, etc.
Gaseous	natural gas	LPG, coal gas, water gas, producer gas, etc.

3.2 CHARACTERISTICS OF A GOOD FUEL

- It should be readily available and cheap.
- Calorific value—It should have a high calorific value.
- Ignition temperature—It should have a moderate ignition temperature. If the ignition temperature is low, the fuel can catch fire easily and the risk of fire hazards is high. If the ignition temperature is very high, it is very difficult to ignite the fuel.
- Low moisture contents—The moisture content of the fuel should be very low because the moisture content reduces the calorific value.
- Controllable combustion—A good fuel should have low non-combustible matter content or ash content because the ash content reduces the calorific value or heating value of the fuel. The disposal of ash is also a big problem and it increases the cost of operation. The combustion should be easily controllable and should not be spontaneously combustible.
- Production of pollution—The products of combustion should not pollute the atmosphere. Gases like CO, SO_2, H_2S, etc. are some of the harmful gases.
- Easy to transport handle and store.
- It should be safe, convenient and economical for storage and transport.

3.3 ADVANTAGES AND DISADVANTAGES OF SOLID, LIQUID AND GASEOUS FUELS

3.3.1 Solid Fuels

Advantages of solid fuels are:

- Solid fuels occur widely and they are cheap.
- They can be handled, stored and transported easily.
- They have a moderate ignition temperature.

Disadvantages of solid fuels are:

- Their calorific value is lower than that of liquid and gaseous fuels.
- They form a lot of ash during burning and disposal of ash is a serious problem.
- The burning process of solid fuels is not as clear as that of liquid and gaseous fuels. Combustion produces smoke, particulate matter, odour and ash.
- A large space is required for storage of solid fuels and sometimes they may undergo spontaneous ignition.
- Since a lot of air is necessary for complete combustion of solid fuels, the thermal efficiency is not so high.

- They cannot be used in internal combustion engines.

3.3.2 Liquid Fuels

Advantages of liquid fuels are as follows:

- They require less storage space than solid fuels.
- As compared to solid fuels, they have high calorific value.
- They can be easily transported over long distances through pipes.
- Liquid fuels burns easily without producing smoke, ash, etc.
- The combustion is uniform and very easily controllable.
- For combustion of liquid fuels, less air is required than that of solid fuels and hence their thermal efficiency is high.
- They can be used in IC engines, boilers and gas turbines.
- They do not undergo spontaneous combustion.

Disadvantages of liquid fuels are as follows:

- Their incomplete combustion causes unpleasant odour.
- In comparison with solid fuels their cost is more.
- Risk of fire hazards is more in case of inflammable and volatile fuels. They should be stored and transported more carefully.
- Some amount of liquid fuels may escape during storage due to evaporation.
- Special type of burners and sprayers are required for effective combustion.

3.3.3 Gaseous Fuels

Followings are the advantages of gaseous fuels:

- Gaseous fuels on burning do not produce any ash or smoke.
- They can be very easily transported any place through pipes and cylinders.
- They have highest calorific value.
- Burning of fuels can be controlled and the nature of flame can be easily made oxidising or reducing by regulating the flow of the fuel.
- They burn without producing smoke, dust and ash.
- Gaseous fuels can be used in IC engines, boilers and gas turbines and in metallurgical operations.
- Compared to solid and liquid fuels, they have high thermal efficiency.

Disadvantages of gaseous fuels are as follows:

- Since gases occupy a large volume, they require large storage tanks.
- Gaseous fuels are highly inflammable and the possibility of fire hazards is very high.
- They are costly.

3.4 CALORIFIC VALUE

The most important property of any fuel is its capacity to supply heat. The efficiency of a fuel is judged by its calorific value. The calorific value of a fuel is defined as the total quantity of heat liberated by the complete combustion of a unit mass or volume of the fuel.

The quantity of heat can be measured by the following units.

(a) **Calorie:** Calorie is defined as the amount of heat required to raise the temperature of 1 gram of water through 1°C.

(b) **Kilocalorie:** It is defined as the quantity of heat required to raise the temperature of 1 kilogram of water through 1°C.

1000 calories = 1 kcalorie

(c) **British Thermal Unit (B.T.U.):** BTU is defined as the quantity of heat required to raise the temperature of 1 pound of water through 1°F.

(d) **Centigrade Heat Unit (C.H.U.):** It is defined as the amount of heat required to raise the temperature of 1 pound of water through 1°C.

1 BThU = 252 calories = 0.252 kcal, 1 kcal = 3.968 BThU = 2.2 CHU

Units of Calorific Value: The calorific values of solid and liquid fuels are expressed in cal/gram or kilocal/kg or BTU/lb. The calorific values of gaseous fuels is expressed in kilocal/m^3.

Gross Calorific Value: When a fuel containing hydrogen is burnt, the hydrogen present in the fuel undergoes combustion and is converted into steam. As the products of combustion are cooled to room temperature, the steam gets condensed into water and latent heat is evolved. Thus, the latent heat of condensation of steam is also included in the calorific value determination which is called as higher calorific value or gross calorific value (GCV).

GCV is defined as the total quantity of heat liberated when a unit mass or volume of the fuel is burnt completely and the products of combustion are cooled to room temperature.

Net Calorific Value: In actual combustion practice, the products of combustion are allowed to escape into the atmosphere with the result that only a lesser amount of heat is available. The amount of heat so available is called lower or net calorific value (NCV). The net calorific value is defined as the quantity of heat produced when a unit mass or volume of the fuel is completely burnt and the products of combustion are allowed to escape.

Net calorific value = Gross calorific value – Latent heat of condensation of water vapour produced

= GCV – Mass of hydrogen per unit weight of the fuel burnt × 9 × Latent heat of condensation of water vapour

One part by weight of hydrogen produces nine parts by weight of water as follows.

$$\begin{array}{ccc} H_2 + O & \longrightarrow & H_2O \\ 2g \quad 16g & & 18g \\ 1 \quad\; 8 & & 9 \end{array}$$

The latent heat of steam is 587 cal/gm.

Thus,

$$NCV = GCV - \frac{9H}{100} \times 587$$

Net calorific value $= GCV - 0.09 \times H \times 587$ cal/gm

where H = % of hydrogen.

3.5 DULONG'S FORMULA

According to Dulong, the calorific value of a fuel is the sum total of calorific values of its constituent elements. The calorific value of carbon, hydrogen and sulphur are 8080, 34,500 and 2240 cal/gm respectively. Nitrogen does not have any calorific value. Oxygen is combined with hydrogen in water. Eight parts by weight of oxygen are combined with 1 part by weight of hydrogen. The hydrogen in moisture is not available for combustion. Thus, the hydrogen available for combustion would be $\left(H - \frac{O}{8}\right)$.

$$\text{Gross calorific value} = \frac{1}{100}\left[8080\,C + 34{,}500\left(H - \frac{O}{8}\right) + 2240\,S\right] \text{ cal/gm}$$

Here C, H, O and S represent the percentage of corresponding elements. These percentages can be found out by ultimate analysis of fuel.

$$\text{Net calorific value} = [G.C.V. - 0.09 \times H \times 587] \text{ cal/gm}$$

By knowing gross and net calorific value by Dulong's formula, we can find out suitability fuel for a specific operation.

Example 3.1: Calculate the higher and lower calorific value of coal having following analysis:

C = 86%, H = 5%, S = 1.5%, N = 2% and O = 3.5%

Latent heat of condensation of water vapour is 587 cal/gm.

Solution:

$$\text{Higher calorific value} = \frac{1}{100}\left[8080\,C + 34{,}500\left(H - \frac{O}{8}\right) + 2240\,S\right] \text{kcal/kg}$$

$$= \frac{1}{100}\left[8080 \times 86 + 34{,}500\left(5 - \frac{3.5}{8}\right) + 2240 \times 15\right]$$

$$= 8556.46 \text{ Kcal/kg}$$

$$\text{Net calorific value} = [\text{G.C.V.} - 0.09 \times \text{H} \times 587]$$

$$= 8556.46 - 264.15 \text{ Kcal/kg} = 8292.31 \text{ Kcal/kg}$$

3.6 SOLID FUELS

Coal

It is believed millions of years ago the earth was covered with thick vegetation, due to earthquake, movements of continents, etc. vegetation got buried in the soil which gradually transformed into coal due to action of heat pressure and bacteria. During this change methane, carbon dioxide and water were main by products. X-ray studies have shown that coal contains clustered, fused aromatic nuclei, stacked one over the other. Coal occurs widely throughout the world, especially in Russia, USA, China, Western Europe and also in India.

The various changes during the transformation are decrease in moisture content, increase in hardness and increase in calorific value.

Classification of coal

Table 3.2 Composition of coals

Type of coal	*%C*	*%H*	*Composition wt. %* *%moisture*	*%Volatile matter*	*Calorific value kcal/kg*
Wood	45–50	4–7	80–95	50–80	4000 to 5000
Peat	45–60	3–6	70–90	45–75	5400 to 5700
Lignite	60–75	4.5–5.5	30–50	45–60	6500 to 7000
Bituminous Coal	75–90	4.0–5.6	1.0–20	11–50	7000 to 8500
Anthracites	92–95	2.9–4.0	1.5–3.5	3.5–10	8500 to 8700

Classification of coal is based on its rank. Wood and peat are soft, lignite is harder and anthracite is a hard with shiny metallic appearance.

The progressive transformation of wood → peat → lignite → bituminous coal → anthracite is associated with increase in rank of coal which is accompanied by:

- Decrease in moisture content, volatile matter content.
- Decrease in percentage of hydrogen, oxygen, nitrogen and sulphur.
- Progressive increase in calorific value.

(1) Peat: Peat is a preliminary stage in the formation of coal. It is brown fibrous soft mass of partially decayed plant material. At the surface aerobic bacteria bring about decay and with increasing depth anaerobic bacteria are responsible for its decay.

Peat is generally used as a fuel for domestic purpose. Due to the high moisture content and low calorific value, it is not used as industrial fuel. It gets powdered during handling which makes its transportation and storage problematic. Low temperature carbonisation of peat gives rise to peat coke, char oil and light spirit. Peat is used as fertiliser. In Russia and Sweden due to shortage of coal, peat is used for generating electricity. In India, peat is found in Nilgiri hills (Tamil Nadu) and in Kolkata on both sides of Hooghly river in West Bengal.

(2) Lignite (Brown coal): Loss of some water will change peat into a still darker and dense rock called lignite which is harder than peat. Lignite should be stored carefully as it absorbs oxygen readily, may ignite spontaneously. Due to its tendency to break into pieces on handling and hygroscopic nature, it is usually used as a local fuel. It burns with long smoky flame. It is used in generating electricity in E. Germany and Russia. In India major deposits are at Neyveli in Tamil Nadu (3800 million tonnes), Nichahom (Kashmir), Kerala coast, Palna and Khari (Rajasthan) and Umarsar (Gujarat) have smaller but significant deposits.

(3) Bituminous coal: It is a further modification of lignite. It burns with yellow smoky flame like bitumen, it is known as bituminous coal. Depending on percentage of carbon, they are divided into two classes: (i) bituminous coal (78-90% carbon), (ii) Sub-bituminous coal (75-83% of carbon).

In India there are extensive bituminous coal deposits in Raniganj, Jharia, Bokaro-Ramgarh-Karnpura, Rajmahal and Santhalpargana (Bihar) Talchar, Hingir-Rampur (Orissa), Singrauli, Korba, Chirimiri, Kanhan valley, Pench valley, Bisrampur (M.P.); Wardh valley, Kamptee-Umrer (Maharashtra) and Singareni and Kottagudem (Andhra Pradesh).

The bituminous coal may have caking ability (ability to form cake on carbonisation) and thus such coals can be used in production of coke. They are known as coking coals. Coals which do not have caking ability are known as non-coking coals.

They are used in generating electricity, steam raising in industrial furnaces, boilers and railway locomotives and for making coke, coal gas as well as for tar, ammonia, benzene, naphthalene and hydrogen sulphide production.

(4) Anthracite: It is the final stage in coal formation. It is the hardest with shiny metallic appearance. It produces short, blue and non-smoky flame. In India it is found in Jammu and West Bengal.

It is used as a domestic fuel, in generating electricity, for naval purposes, central heating furnaces, for generating producer gas and for curing rubber.

3.7 SELECTION OF COAL

Various properties of coal which determine selection of coal for different applications.

- **Calorific value:** Calorific value of the coal should be high. A coal with higher calorific value is usually preferred.
- **Moisture content:** A coal with low moisture content has better calorific value and less expensive and thus more preferred. The moisture content should be as low as possible.
- **Ash content:** Ash is a non-combustible inorganic matter in coal which reduces calorific value of the coal. Coal which produces low ash is usually preferred.
- **Calorific intensity and flexibility:** During combustion of fuel, the area of flame depends on its volatile matter content. More the volatile matter content, larger is the area of flame. If area of the flame is larger, concentration of heat is lesser and thus calorific intensity is also less. A coal with high calorific intensity is preferred.
- **Size of coal:** A coal with uniform size can be stored and handled more easily.
- **Sulphur and phosphorus content:** Very low sulphur and phosphorus content in coal is essential in order to avoid pollution of atmosphere and production of solid matter during combustion.

3.8 ANALYSIS OF COAL

Coal is composed of carbon, hydrogen, oxygen, nitrogen, sulphur, etc. along with some inorganic matter. The calorific value of coal depends upon its composition. In order to determine the quality of coal, two types of analyses are carried out: (A) Proximate analysis (B) Ultimate analysis

3.8.1 Proximate Analysis

This method is used for determining moisture content, volatile matter, ash content and fixed carbon in the coal. It provides important information regarding the suitability of coal for a particular domestic or industrial use. It is carried out as per standard specifications, by using air-dried sample.

(1) Moisture content

The moisture of air dried, finely powdered coal sample is determined by heating accurately weighed quantity of a sample in a silica crucible at 105-110°C for 1 hour in an electric oven. The loss in weight of sample is found out by the difference in weights. The percentage of moisture is calculated as

$$\text{Percentage of moisture} = \frac{\text{Loss in weight of sample}}{\text{Weight of air} - \text{dried sample}} \times 100$$

Significance

During combustion a significant amount of heat is wasted in evaporating the moisture. High percentage of moisture reduces the calorific value of coal. As moisture is transported at the same cost as coal, fuel becomes expensive. Excessive surface moisture is also troublesome in handling of coal. However, low percentage of moisture is desirable in coal as it improves calorific value and reduces formation of ash.

(2) Volatile matter

The percentage of volatile matter is determined by heating accurately weighed air-dried coal sample in a silica crucible covered with lid. The heating is carried out at 925 ± 5°C in a muffle furnace for exactly seven minutes in the absence of air.

$$\text{Percentage volatile matter} = \frac{\text{Loss in weight of sample}}{\text{Weight of air} - \text{dried sample}} \times 100 - \text{\% of moisture}$$

Significance

The volatile matter evolved consists of combustible gases like CO, H_2, CH_4 and other hydrocarbons and incombustible gases like CO_2 and N_2. The incombustible gases do not have any heating value and are undesirable. Some coals may have high volatile matter (upto 50%) and on burning they produce a large amount of gas or vapour which also burn with production of flame. Thus, coals with high volatile matter burns with long, smoky flame and coals with low volatile matter burn with a shorter flame. Thus, the information regarding the percentage of volatile matter is essential while designing a furnace, as combustion space required depends upon the nature of flame.

The caking ability (coking power) varies with the percentage of volatile matter. The coals having volatile matter up to 20 to 30% are more suitable for production of metallurgical coke.

(3) Ash content

A coal sample after determination of volatile matter is then heated in a muffle furnace at 725-750°C without lid for 1 hr. cooled and weighed. Process of ignition, cooling and weighing is repeated till combustion is complete which is indicated by constant weight of the residue.

$$\text{Percentage of ash} = \frac{\text{Weight of ash (residue)}}{\text{Weight of dried sample}} \times 100$$

Significance

Ash is a non-combustible inorganic matter consists of silica, alumina, iron oxide and small quantities of magnesia as well as lime.

The ash reduces calorific value of coal. It increases transportation and storage cost of coal. Since ash has to be removed from the furnace and disposed of, it increases labour cost.

As coals are used as reducing agents in several metallurgical operations, the composition of ash affects the slag and metal composition, its characteristics and it is the most important factor governing the selection of the proper flux. Ash results in the formation of clinkers (lumps of ash) which causes uneven burning due to interruption of air supply.

(4) Fixed Carbon

The fixed carbon in the coal can be calculated as,

Percentage of fixed carbon = 100 – (percentage moisture + percentage volatile matter + percentage ash)

Quality of coal depends on the percentage of fixed carbon, it should be as high as possible. Fixed carbon varies inversely with volatile matter. It represents the amount of carbon that will burn in solid state. More the Percentage of fixed carbon, higher is the calorific value.

Example 3.2: 2.5 g of air-dried coal sample was taken in a silica crucible, after heating it in an electric oven at 105-110°C for 1 hour, the residue was weighed 2.415 g. The residue was then ignited at 725-750°C to a constant weight of 0.2469 g.

In another experiment, 1 g of the same sample was heated in a silica crucible covered with vented lid at a temperature 950 ± 20°C for exactly 7 minutes. After cooling the weight of residue was found to be 0.611 g. Calculate the percentage of fixed carbon.

Solution: (1) Percentage of moisture:

$$\left[\begin{array}{l}\text{Loss in weight when heated at } 105-110^\circ\text{C} \\ \text{for 1 hr (mass of moisture in coal)}\end{array}\right] = 2.500 - 2.415 = 0.085 \text{ g}$$

$$\text{Percentage of moisture} = \frac{\text{Loss in weight}}{\text{Weight of sample}} \times 100$$

$$= \frac{0.085}{2.5} \times 100 = 3.4$$

(2) Percentage of ash:

Weight of residue after ignition at 700-750°C = 0.246 g

$$\text{Percentage of ash} = \frac{\text{Weight of residue}}{\text{Weight of sample}} \times 100$$

$$= \frac{0.2469}{2.5} \times 100 = 9.87$$

(3) Percentage of volatile matter:

$$\left[\begin{array}{l}\text{Loss in weight on heating in silica crucible} \\ \text{with vented lid at } 950 \pm 20°\text{C for 7 min}\end{array}\right]$$

$$= 1.000 - 0.611 = 0.389 \text{ g}$$

$$\text{Percentage of V.M} = \frac{\text{Loss in weight}}{\text{Weight of sample}} \times 100 - \%\text{moisture}$$

$$= \frac{0.389}{1} \times 100 - 3.40$$

$$= 38.90 - 3.40 = 35.50$$

(4) Percentage of fixed carbon = 100 – % [Moisture + Volatile matter + Ash]

$$= 100 - [3.40 + 35.50 + 9.87]$$

$$= 100 - 48.77$$

$$= 51.23$$

3.8.2 Ultimate Analysis of Coal

The proximate analysis gives an idea of gross impurities present. The ultimate analysis gives an exact amount of elements such as carbon, hydrogen, sulphur, nitrogen, oxygen and ash.

(1) Determination of carbon and hydrogen

Accurately weighed quantity of air-dried coal sample is burnt in stream of oxygen at 1250°C in a combustion tube. The carbon and hydrogen in coal get oxidised to CO_2 and H_2O respectively.

$$\underset{2}{C} + \underset{32}{O_2} \longrightarrow \underset{44}{CO_2}$$

$$\underset{2}{H_2} + \underset{16}{\frac{1}{2}O_2} \longrightarrow \underset{18}{H_2O}$$

The products of combustion are passed through a water absorbing unit containing anhydrous magnesium perchlorate or calcium chloride and then through carbon dioxide absorbing unit containing soda-lime or potassium hydroxide. Increase in weight of these units is found out which corresponds to weight of H_2O and CO_2 formed due to combustion of carbon and hydrogen respectively.

$$\text{Percentage of hydrogen} = \frac{2}{18} \times \frac{\text{Increase in weight of water absorbing unit}}{\text{Weight of coal sample}} \times 100$$

$$\text{Percentage of carbon} = \frac{12}{44} \times \frac{\text{Increase in weight of } CO_2 \text{ absorbing unit}}{\text{Weight of coal sample}} \times 100$$

Example 3.3: 0.250 g of coal sample was burnt in stream of oxygen at 1290°C in a combustion tube. There was increase in weight of magnesium perchlorate unit by 0.025 g and that of soda-lime unit by 800 mg after combustion. Calculate % of carbon and hydrogen of coal sample.

Solution: Increase in weight of magnesium perchlorate unit = Weight of water formed due to combustion of hydrogen = 0.025 g

Thus,
$$\text{Percentage H} = \frac{2}{18} \times \frac{\text{Weight of } H_2O \text{ formed}}{\text{Weight of sample}} \times 100$$

$$= \frac{2}{18} \times \frac{0.025}{0.250} \times 100 = 1.11$$

Increases in weight of soda-lime unit = Weight of CO_2 formed due to combustion of carbon = 0.800 g

Thus,
$$\text{Percentage C} = \frac{12}{44} \times \frac{\text{Weight of } CO_2 \text{ formed}}{\text{Weight of sample}} \times 100$$

$$= \frac{12}{44} \times \frac{0.800}{0.250} \times 100 = 87.27$$

Significance of carbon and hydrogen

Carbon is a major constituent of coal. More the percentage of carbon, higher is the calorific value of the coal. Percentage of carbon is useful in assessing the quality of coal.

The percentage of hydrogen in coal sample reflects greatly in its calorific value, calorific value of hydrogen is four times more than that of carbon. Hydrogen is mostly associated with volatile matter of coal and therefore is a deciding factor in the design of combustion chamber.

(2) Determination of nitrogen

Nitrogen does not contribute towards calorific value. It is determined by Kjeldahl's method. In this method, coal sample is heated with conc. H_2SO_4 in the presence of a catalyst like selenium powder or a mixture of mercuric sulphate and potassium sulphate. Nitrogen in coal thus gets converted to ammonium sulphate and a clear solution is obtained.

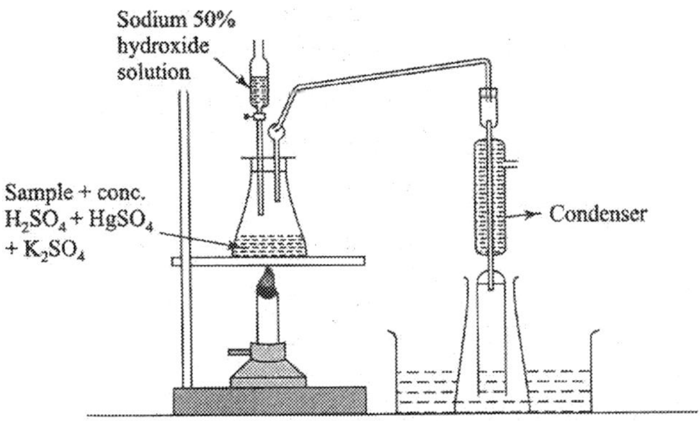

Fig. 3.1 Ammonia distillation apparatus.

$$2N + 6H + H_2SO_4 \longrightarrow (NH_4)_2\,SO_4$$

The solution is then treated with 50% NaOH, to liberate ammonia.

$$(NH_4)_2\,SO_4 + 2NaOH \longrightarrow 2NH_3\uparrow + Na_2SO_4 + 2H_2O$$

The liberated ammonia is then distilled over and absorbed in a known quantity of standard sulphuric acid.

$$2NH_3 + H_2SO_4 \longrightarrow (NH_4)_2\,SO_4$$

Depending upon the amount of liberated ammonia, equivalent amount of H_2SO_4 will be utilised to absorb it and unused quantity of H_2SO_4 is determined in terms of N/10 NaOH in a back titration (V_2 ml). A blank titration is carried out to know total quantity of H_2SO_4 in terms of N/10 NaOH. In blank titration, same quantity of std. H_2SO_4 used to absorb ammonia, is titrated against N/10 NaOH (V_1 ml). The difference between blank and back titration gives amount of H_2SO_4 in terms of N/10 NaOH, utilised to absorb ammonia ($V_1 - V_2$ ml).

$$2NaOH + H_2SO_4 \longrightarrow Na_2SO_4 + 2H_2O$$

From the above equation, we can write equivalence as

$$\begin{aligned} 2NaOH &\equiv H_2SO_4 \equiv 2NH_3 \equiv (NH_4)_2\,SO_4 \equiv 2N \\ NaOH &\equiv N \end{aligned}$$

Thus, 1000 ml of 1 N NaOH will correspond to 14 g of nitrogen in coal. 1000 ml of 0.1 N will correspond to 1.4 g of nitrogen. The percentage nitrogen in coal can be calculated by following equation, where W is weight of coal sample.

$$\text{Percentage N} = \frac{(V_1 - V_2) \times \text{Normality of NaOH} \times 1.4}{W}$$

Example 3.4: In estimation of nitrogen, 1.5 g of coal sample liberated ammonia which was absorbed in 50 ml of 0.5 N H_2SO_4. The resultant solution required 12.5

ml of 0.1 N NaOH solution for complete neutralisation in a back titration. A blank titration reading was 25 ml. What is percentage of nitrogen in coal?

Solution:

$$V_1 \text{ ml} = \text{Blank titration reading} = 25 \text{ ml}$$

$$V_2 \text{ ml} = \text{Back titration reading} = 12.5 \text{ ml}$$

$$(V_1 - V_2) \text{ ml} = (25 - 12.5) \text{ ml} = 12.5 \text{ ml}$$

$$W = \text{Weight of sample} = 1.5 \text{ g}$$

$$\text{Percentage N} = \frac{(V_1 - V_2) \times 0.1 \times 1.4}{W}$$

$$= \frac{12.5 \times 0.14}{1.5}$$

$$= 1.166$$

Significance of nitrogen

Combustion of nitrogen at high temperature (1000°C) produce nitrogen oxides (NO_x) which escape with flue gases and cause atmospheric pollution NO_x on contact with water and ozone form nitric acids as a result of photochemical reaction in atmosphere.

$$NO + O_3 \longrightarrow NO_2 + O_2$$

$$NO_2 + O_2 \longrightarrow NO_3 + O_2$$

$$NO_2 + NO_3 \longrightarrow N_2O_5$$

$$N_2O_5 + H_2O \longrightarrow 2HNO_3$$

Nitrogen has no calorific value. Hence, its presence is not desirable in coal.

(3) Determination of sulphur

In the estimation of sulphur, washing from bomb calorimeter in finding out calorific value are used. Sulphur present in coal gets converted to sulphur dioxide which on cooling in calorimeter is absorbed by water to form dilute sulphuric acid. Sulphate is precipitated by treatment with $Ba(NO_3)_2$ solution precipitated $BaSO_4$ is filtered hot, washed dried and ignited to constant weight.

$$SO_2 + H_2O \longrightarrow H_2SO_3$$

$$2H_2SO_3 + O_2 \longrightarrow 2H_2SO_4$$

$$H_2SO_4 + Ba(NO_3)_2 \longrightarrow BaSO_4\downarrow + 2HNO_3$$

Significance

Presence of sulphur is undesirable it produces SO_2 and H_2SO_3 which is corrosive.

(4) Determination of ash

Ash is determined in the same way as in proximate analysis.

(5) Determination of oxygen

Percentage of oxygen is calculated as follows:

Percentage of Oxygen = 100 – [percentage C + percentage H + percentage N + percentage S + percentage Ash]

Significance

Higher percentage due to mixture constant which reduces calorific value. For a good quality of coal should have very low oxygen constent.

Example 3.5: A coal sample was subjected to ultimate analysis, following results are obtained.

(i) 0.24 g of sample on combustion gave 0.792 g of CO_2 and 0.0216 g H_2O.

(ii) 1.4 g of sample in estimation of nitrogen by a Kjeldahl's method liberated ammonia was absorbed in 50.0 ml 0.1 N H_2SO_4. The resultant solution required 10 ml of 0.1 N NaOH for complete neutralisation. The blank titration reading is 50 ml.

(iii) 3.2 g of sample in determination of sulphur gave 0.233 g $BaSO_4$.

(a) Using Dulong formula, calculate gross and net calorific value of the coal.

(b) Calculate the amount of O_2 and air required for complete combustion of 1 kg of coal.

(c) Also calculate volume of O_2 and air at 750 m pressure of 25°C.

Solution: (i) Percentage of carbon and % of hydrogen.

$$\text{Percentage carbon} = \frac{12}{44} \times \frac{\text{Weight of } CO_2 \text{ formed}}{\text{Weight of sample}} \times 100$$

$$= \frac{12}{44} \times \frac{0.792}{0.24} \times 100 = 90$$

$$\text{Percentage hydrogen} = \frac{2}{18} \times \frac{\text{Weight of } H_2O \text{ formed}}{\text{Weight of sample}} \times 100$$

$$= \frac{2}{18} \times \frac{0.0216}{0.24} \times 100 = 1$$

(ii) Percentage of nitrogen

(1) Back titration required 10 ml of 0.1 N NaOH. It corresponds to unused quantity of NaOH-V_2 ml.

(2) Blank titration result – As 50.0 ml 0.1 N H_2SO_4 is utilised to absorb ammonia, same quantity must be titrated with 0.1 N NaOH.

$$\begin{aligned} \text{NaOH} \quad & \quad H_2SO4 \\ N_1V_1 &= N_2V_2 \\ 0.1 \times V_1 &= 0.1 \times 50 \\ V_1 &= \frac{1}{10} \times 50 \times 10 = 50 \text{ ml} \end{aligned}$$

The blank titration reading would be 50 ml. Thus, $V_1 = 50$ ml

$$\begin{aligned} (V_1 - V_2) \text{ ml} &= 50 - 10 = 40 \text{ ml} \\ \text{Weight of sample} &= W = 1.4 \text{ g} \\ \text{Percentage N} &= \frac{(V_1 - V_2) \times 0.1 \times 1.4}{W} = \frac{40 \times 0.14}{1.4} = 4 \end{aligned}$$

(3) Percentage of sulphur in coal sample

$$\begin{aligned} \text{Weight of sample} &= 3.2 \text{ g} \\ \text{Weight of } BaSO_4 &= 0.233 \text{ g} \\ \text{Percentage S} &= \frac{32}{233} \times \frac{\text{Weight of } BaSO_4 \text{ formed}}{\text{Weight of sample}} \times 100 \\ &= \frac{32}{233} \times \frac{0.233}{3.2} \times 100 = 1 \end{aligned}$$

(4) Percentage of O then can be calculated as

$$\begin{aligned} \text{Percentage of O} &= 100 - \%\,[C + H + N + S + Ash] \\ &= 100 - [90 + 1 + 4 + 1 + 0] = 100 - 96 = 4 \end{aligned}$$

The gross calorific value can be calculated by Dulong's formula

$$\begin{aligned} \text{G.C.V.} &= \frac{1}{100}\left[8080\,C + 34500\left(H - \frac{O}{8}\right) + 2240\,S\right] \\ &= \frac{1}{100}\left[8080 \times 90 + 34500\left(1 - \frac{4}{8}\right) + 2240 \times 1\right] \\ &= \frac{1}{100}[727200 + 17250 + 2240] \\ &= 7466.9 \text{ cal/gm} \\ \text{Net calorific value} &= \text{G. C. V} - 0.09 \times H \times 587 \\ &= 7466.9 - 0.09 \times 1 \times 587 \\ &= 7466.9 - 52.83 \\ &= 7414.07 \text{ cal/gm} \end{aligned}$$

3.9 LIQUID FUELS

They are very important as they are used in almost all internal combuation engines and generating heat in ovens and furnaces.

3.9.1 Petroleum

Petroleum is a natural product generated from organic material by physical and chemical processes inside the earth. The term petroleum means oil obtained from rock. It is also known as crude oil, mineral oil, crude or simply oil. Crude petroleum is formed from buried vegetable and marine life due to the action of heat pressure and microbial activity over millions of years.

The major oil fields (as 56.3%) in the world are located in middle east (mainly Saudi Arabia, Iran, Kuwait, Iraq and Abu Dhabi). Other major oil bearing countries are Russia, Mexico, USA, Venezuela, Libiya and Nigeria.

Indian oil fields are mainly in Assam, Gujarat and Maharashtra.

3.9.2 Composition

Petroleum oil occurs in nature as brownish green to black coloured viscous liquid. It consists of a mixture of hydrocarbons along with small quantities of organo-metallic compounds of iron, vanadium, copper, nickel, etc.

Composition of crude oil (ultimate analysis):

Percentage of carbon:	79 to 86%
Percentage of hydrogen:	11.0 to 15.0%
Percentage of oxygen:	0.1 to 9.0%
Percentage of sulphur:	0.1 to 5%
Percentage of nitrogen:	0.4 to 0.9%

The crude oil contains following types of hydrocarbons:

(1) **Paraffins/Alkanes:** They have general formula C_nH_{2n+2}. $C_1 - C_4$ alkanes are in gaseous state, $C_5 - C_{17}$ are liquids and C_{18} onwards are in solid state. Straight chain as well as branched chain alkanes are present in crude oil. The proportion of different fractions are as four parts of diphatic hydrocarbon + one part of aromatic + one part of naphthalene.

(2) **Naphthenes/Cycloalkanes/Cycloparaffines:** They are saturated cyclic compounds having general formula C_nH_{2n}.

(3) Aromatics e.g. benzene, toluene, xylene, etc.

(4) Olefinic hydrocarbons are absent in crude oil.

(5) Sulphur, nitrogen and oxygen are present as derivatives of hydrocarbons.

According to Kuldeep Chandra et al., (1989) the composition of average crude of 35 API Gravity is as follows:

Table 3.3

Fraction	*Composition*	*Volume percentage*
Gasoline	$C_5 - C_{10}$	27
Kerosene	$C_{10} - C_{18}$	13
Diesel	$C_{13} - C_{20}$	12
Heavy gas oil	$C_{19} - C_{25}$	10
Lubricating oils	$C_{26} - C_{40}$	20
Residue	C_{40}	18

3.9.3 Classification of Crude Oil on the Basis of their Composition

The classification of crude oil in terms of composition is based on amount of alkane (paraffins), cycloalkane (naphthenes) and aromatic compounds in oil along with N, S, O compounds, i.e. resins and asphaltenes.

(1) **Paraffinic base crude oil:** They contain normal and isoalkanes and less than 1% S. The asphaltene and resin content is below 10%. Percentage of aromatic hydrocarbons is low. They yield gasoline of low octane number and gas oil of high cetane value. They give high grade lubricating oils on distillation and leave behind paraffin wax as residue.

(2) **Naphthenic base crude oil:** They contain cycloalkanes (naphthenes). Asphaltic matter is present in quite large proportion. Paraffinic wax is very little or absent. These crude oils yield good quality gasolines and asphalts/bitumens.

(3) **Paraffinic-Naphthenic base crude oils/mixed base crude oil:** They have both paraffinic and naphthenic hydrocarbons together with a certain proportion of aromatic hydrocarbons. On distillation, they yield residue containing both parafin waxes and asphaltic matter, their density and viscosity is more than paraffinic class.

3.9.4 Oil Industry

Petroleum oil is a major source of world's energy and chemicals.

(1) Petroleum Industry

It usually involves the following operations:

It involves various operations for locating petroleum deposits below the earth surface, drilling of deep wells to tap oil reservoirs, operations to bring oil and gas to the surface, either by self energy or by pumping, etc., separation of crude oil from gas, water and sand and addition of flow improvers to it.

The crude oil is then transported from oil field to refinery site.

(2) Petrochemical Industry

It usually involves fractional distillation of crude oil to segregate various fractions (useful products), subsequent processing steps such as cracking, alkylation, catalytic reforming, hydroprocessing, etc. to increase the yield of products which have strong market demand or to improve the performance of some of the products.

3.9.5 Mining of Petroleum/Production of Oil

Oil Reservoir

An oil reservoir can occur in nature in a variety of forms, depending upon the geometry of the trap. However, oil and gas present in an anticline can be depicted by simple illustration.

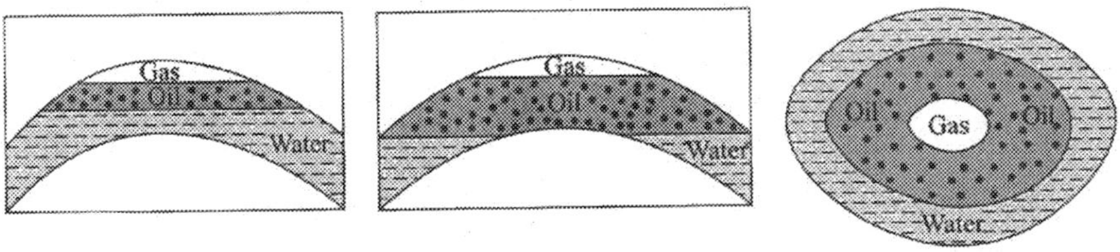

(a) Bottom water (b) Edge water (c) Plan of field 'A'

Fig. 3.2 Bottom-water and edge-water.

Primary and Secondary Recovery

Primary recovery

Primary recovery is the recovery of oil without any external aid that is allowing the oil to come out on its own reservoir energy. Control at this stage involves regulating the rate of flow.

Secondary recovery

Secondary recovery is the recovery of oil with some external aid. The most important secondary method for maintaining the pressure of a formation is the repressuring. This is done by injecting water below the oil-water contact. Figs. 3.3 shows mechanism of injection. The quality and quantity of water injected has also to be monitored. If water contains clay particles which will flocculate in the formation and reduce the injectivity. Similarly, water-containing chemicals cause complications. The water must be available in large quantity as thousands of tonnes of oil is recovered per day, one and a half times that amount of water must be available for injection. The well, through which water is injected is known as injection well and the well through which oil is recovered, is known as producing well. This method is known as water flooding.

It is also possible to inject gas in the gas-cap to prop up the depleting pressure. Gas which comes out from the same oil can also be used.

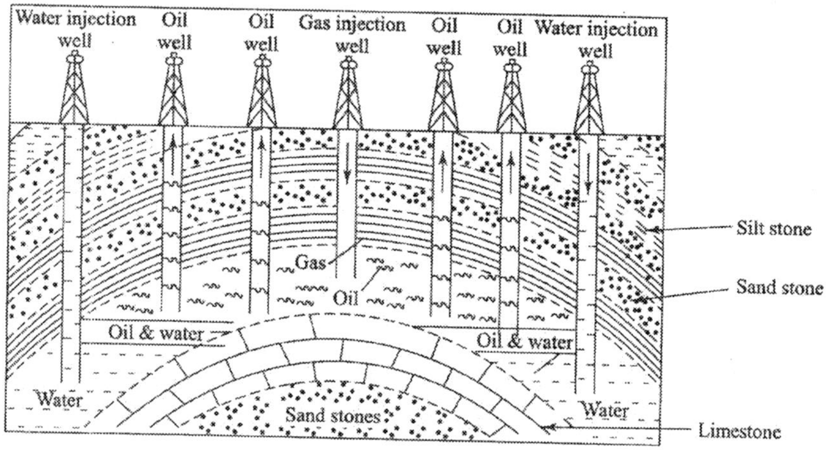

Fig. 3.3 Water and gas injection and production of oil.

Enriched hydrocarbon gases and CO_2 can also be used as a medium of miscible displacement to improve volumetric sweep of oil. 20-30% production as compared to water flooding can be improved by fire-flood method.

In this method, viscosity of oil is reduced by application of heat. For this air is injected into reservoir and is ignited by a heater or a spark put against the formation inside the well. Once the fire-flood starts in the formation, it is necessary to keep on injecting air to sustain fire-flood. Steam or hot water is also injected into the well, to reduce viscosity of oil.

3.9.6 Refining of Crude Oil (Fractional distillation)

Crude oil obtained from earth as a result which contains impurities. It is necessary to break the emulsion and purify before fractional distillation.

Demulsification

Crude oil occurs as a emulsion, this emulsion is broken by addition of demulsifying agent, like $CaCl_2$, NaCl, $MgCl_2$. It can be are also removed by repeated washings with water.

Removal of sulphur

Undesirable elements present in crude oil are removed by treatment with copper oxide. For e.g. Insoluble black copper sulphide is removed by filtration.

$$CuO + S \longrightarrow CuS\downarrow$$

Fractional distillation

Crude oil is heated at 400°C and vapours are passed through fractionating column. In fractional distillation, each individual vaporisation, condensation and reboiling is carried out in fractionation towers. This process enables us to obtain a few major fractions at different temperature ranges. The composition and uses of main petroleum fractions are tabulated in Table 3.4.

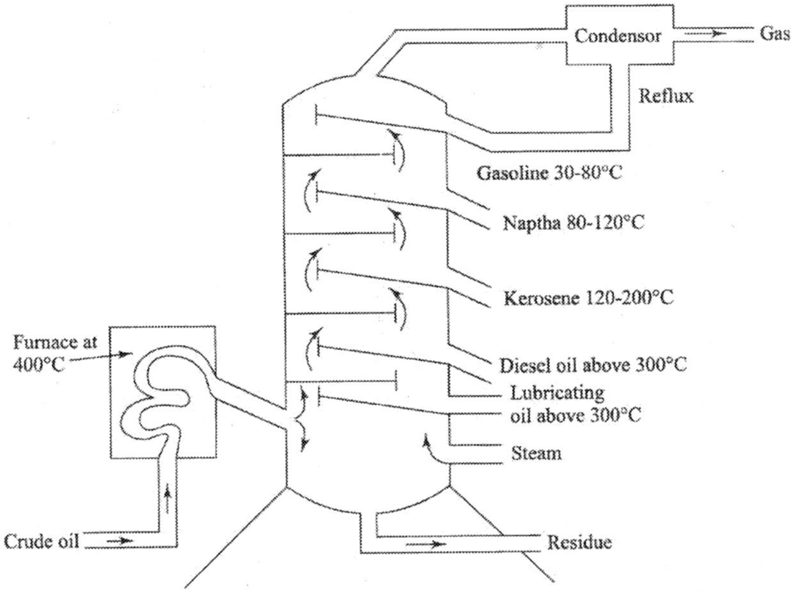

Fig. 3.4 Fractionating column.

The significant operations in petroleum refineries are:

1. To produce high octane gasoline for automobile.
2. The preparation of hydrocarbon feed-stock for petrochemicals manufacture.

Fractional distillation allows the separation of crude oil into a variety of useful products. Often, subsequent processing steps are necessary to increase the yield of product which have strong market demand or to improve the performance of some of the products. Some of these processes/refinery operations are cracking, reforming, hydroprocessing, alkylation, etc.

3.9.7 Cracking

Cracking of hydrocarbons is a process of breaking of large high molecular weight hydrocarbons into smaller hydrocarbons and olefins at high temperature known as thermal cracking or at low temperature in the presence of catalyst called catalytic cracking. The product of cracking of alkanes depends upon the following factors.

Table 3.4 The Composition and Uses of Main Petroleum Fractions

Name of fraction	*Composition*	*Temperature range*	*Uses*
(1) Natural gas	$C_1 - C_4$	Below 40°C	LPG domestic fuel and as a source of many organic chemicals
(2) Petrol or gasoline on further fractionation gives	$C_5 - C_{10}$	30–200°C	
(i) Petroleum ether	$C_5 - C_6$	30–80°C	Widely used as solvent
(ii) Gasoline or petrol (straight run gasoline)	$C_5 - C_9$	80–120°C	Motor fuel, solvent
(iii) Naptha	$C_9 - C_{10}$	120–180°C	As a solvent for paints and varnishes
(3) Kerosene	$C_{10} - C_{18}$	180–250°C	Domestic fuel, aviation fuel in jet engine, as illuminant.
(4) Heavy oil on further fractionation gives	$C_{16} - C_{30}$	200–350°	
(i) Fuel oil		Above 200°C	As a fuel
(ii) Diesel oil		Above 300°C	
(iii) Heavy oil	$C_{17} - C_{30}$		Fuel for ship and for cracking conversion to gasoline
(5) Residual oil on fractionation gives	$C_{18} - C_{40}$		
(i) Lubricating oil	$C_{18} - C_{20}$		As lubricant
(ii) Paraffin wax	$C_{20} - C_{28}$		Candles, bootpolish, wax, papers, insulation
(iii) Petroleum jelly	$C_{20} - C_{30}$		Medicines, cosmetics, lubricants
(6) Residue			
(i) Asphalt			Used for water proofing of roofs, paying playgrounds and roads.
(ii) Petroleum coke			Used as fuel.

(i) Structure of alkane

(ii) Pressure employed

(iii) Presence or absence of catalyst

The primary reactions during cracking are represented as:

[High Mol wt hydrocarbons having high B.P. and cannot be used as fuel] $\xrightarrow{\text{Cracking}}$ [Low Mol. wt hydrocarbons having low B.P., more volatility and can be used as fuel]

For example, dodecane ($C_{12}H_{26}$) is a big hydrocarbon present in heavy oil fraction. Its boiling point is 216°C due to which it cannot be used as petrol. On cracking it gives smaller hydrocarbon molecules, heptane and pentene having boiling point 98°C and 36°C respectively which can be used as petrol.

$C_{12}H_{26}$		C_7H_{16}		C_5H_{10}
Dodecane	$\xrightarrow{\text{Cracking}}$	Heptane	+	Pentene
Boiling point		Boiling point		Boiling point
216°C		98°C		36°C

Another example

$C_{10}H_{22}$	$\longrightarrow$	C_5H_{12}	+	C_5H_{10}
Decane		*n*-pentane	+	Pentene
(B.P. 174°C)				(36°C)

On fractionation of crude oil 20-30% gasoline is obtained which is not adequate to meet its demand. Thus, less desirable petroleum fractions having more than 9 carbon atoms are cracked to $C_5 - C_9$ range hydrocarbons which are important components of gasoline. Straight run gasoline has octane as low as ~ 30 and the yield is 20%. By thermal cracking yield is increased to 40% and octane number ~ 70. In catalytic cracking, the yield is increased to 50% and octane number to ~ 100.

The different cracking methods are given below.

Thermal cracking

When cracking is carried out at high temperature without any catalyst, it is known as thermal cracking. There are two types of thermal cracking.

Liquid phase thermal cracking: In this method, the heavy oil is cracked at 475-530°C under high temperature (100 kg/cm^2) to keep the reaction product in liquid state. The cracked products are then separated in a fractionating column. This method gives about 50-60% yield.

Vapour phase thermal cracking: In this method, the cracking oil is vaporised and then cracked at 600-650°C under a pressure 10 to 20 kg/cm^2. The yield of petrol is about 70 percent. This process is suitable only for those oils which are readily vaporised. Heavy oils cannot be cracked by this method because these oils cannot be completely vaporised under these conditions.

Catalytic cracking

This type of cracking is carried out at much lower temperature and pressure in the presence of a catalyst. The catalysts used are hydrated aluminosilicates or zeolites. There are two types of catalytic cracking.

Fixed bed catalytic cracking: The heavy oil is passed through a heater at 400-500°C. The bauxite catalyst is mixed with clay and zirconium oxide and packed in catalyst chamber. The hot vapours of oil are passed over a fixed bed of catalyst in the catalyst chamber maintained at 425-450°C and 1.5 kg/cm^2 pressure. During this passage through the chamber, about 40 per cent of the heavy oil charge is converted into gasoline. About 2-4 percent carbon formed, gets absorbed on the catalyst bed. The cracked vapours are then passed through the fractionating column to condense heavy fractions. The vapours are then passed through the cooler where gasoline along with some gases. The dissolved gases are removed and pure gasoline fraction is recovered by passing through the degasifier. Due to the deposition of carbon the catalyst after 8-10 hours stops its functioning. It is reactivated by burning off the carbon deposited in a steam of hot air.

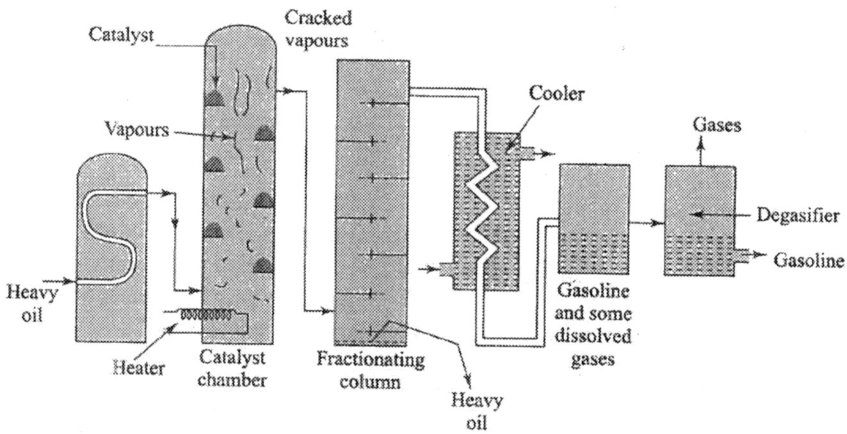

Fig. 3.5 Fixed bed catalytic cracking.

Moving bed or Fluid bed catalytic cracking: In this process, very finely powered catalyst is circulated in gas stream which behaves almost as a fluid.

The oil vapours are cracked in a reactor maintained at 530°C and pressure of 5 kg/cm^2. The catalyst dust is removed by passing the cracked vapours through the catalyst dust separator. The cracked vapours are then passed through the fractionating column where heavy oil is separated. Then vapours are passed through the cooler where gasoline condenses along with some dissolved gases. The dissolved gases are further removed by passing the gasoline through the degasifier.

The yields of gasoline by cracking catalytic process is around 40% of feed stock and has octane rating of 70-80.

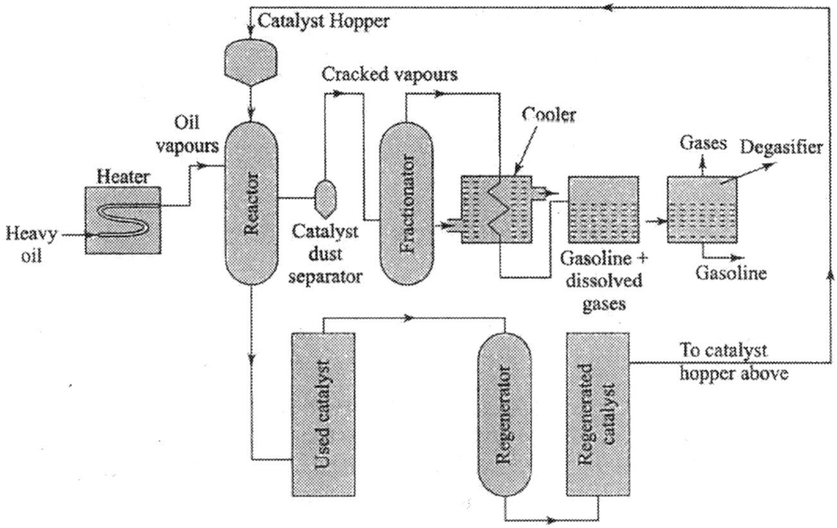

Fig. 3.6 Moving bed catalytic cracking.

Advantages of catalytic method

- Temperature and pressure used are lower as compared to thermal cracking method.
- Catalysts are specific in their action, they only break down the high mol. wt. hydrocarbons.
- The stability of gasoline is better due to complete removal of dissolved unsaturated hydrocarbons.

3.9.8 Knocking

In I.C. engine, petrol is used as fuel. Two important characteristics of petrol are its volatility and antiturbulence. In an I.C. engine, fuel must remain liquid until it enters the air stream in the carburettor inlet. At this point it should vaporise and mix with the air. The mixture of petrol vapour and air is highly compressed before it is ignited by a spark to have maximum efficiency. Under certain conditions there may be spontaneous combustion before it is ignited by the spark, which may lead to explosive burning of petrol and a characteristic metallic sound called knocking. This results in wastage of fuel and damage to the engine. The knocking tendency of gasoline depends on the structure of hydrocarbons present in it. The straight chain paraffins have more knocking tendency than branched chain paraffins. Small sized naphthalenes have medium knocking tendency and aromatics such as benzene have less knocking tendency. Straight chains > Branched chain > Cycloalkanes > Alkenes > Bromatics > Hydrocarbons > Alkanes. The knocking tendency of gasoline can be measured in terms of its octane number. More the octane number, less is the knocking tendency. The iso-octane (2, 2, 4 – trimethyl pentane), a highly branched paraffin, has very good knocking resistance. It is given an arbitrary octane number 100 and the n-heptane which has poor resistance to knocking due to its straight chain structure, is given an arbitrary octane number 0. Higher the octane number, better is the quality of fuel.

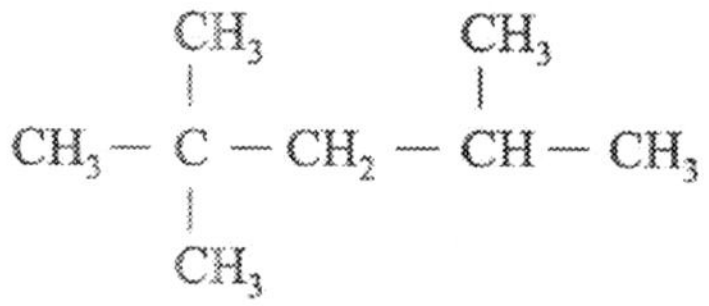

$CH_3-(CH_2)_5-CH_3-$

Iso-octane, octane number-100

n-Heptane, octane number-Zero

The octane number of gasoline is defined as the % of iso-octane present in the mixture of iso-octane and *n*-heptane which matches with gasoline in knocking, e.g. gasoline is said to have octane number 80 when it matches its knocking property with mixture of 80% iso-octane and 20% n-heptane. A suitable proportion of lower boiling paraffins must be present to facilitate cold engine start in petrol.

Table 3.5 Octane numbers of some hydrocarbons

Name	Structure	Octane number
n-Heptane	$CH_3-(CH_2)_5-CH_3$	00
3-Methyl hexane	$CH_3-CH_2-CH(CH_3)-CH_2-CH_2-CH_3$	52
3-Ethypentane	$CH_3-CH_2-CH(CH_2-CH_3)-CH_2-CH_3$	65
2,3-Dimethyl pentane	$CH_3-CH(CH_3)-CH(CH_3)-CH_2-CH_3$	75
2,2,4-Trimethyl pentane (Iso-octane)	$CH_3-C(CH_3)_2-CH_2-CH(CH_3)-CH_3$	100
Benzene	C_6H_6 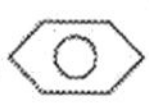	106
Toluene	$C_6H_5-CH_3$ 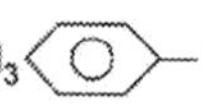 CH_3	118

Antiknocking agents

The octane rating of many fuels can be increased by the addition (doping) of certain organometallic compounds to gasoline. The substances added to motor or aviation gasoline for controlling and containing the knocking in I.C. engines are called 'antiknocking' agents.

The common commercial antiknocking agents used are:

Tetraethyl lead (TEL) $Pb(C_2H_5)_4$, Tetramethyl lead (TME) Pb $(CH_3)_4$, Mixed methylethyl lead.

TEL is most widely used as it is cheap and more effective in increasing the octane rating of fuels. It is normally used along with ethylene dibromide or ethylene dichloride to convert less volatile Pb and PbO into more volatile $PbBr_2$ or $PbCl_2$.

The knocking is due to free radical reactions taking place at low temperature. Knocking can be prevented by adding small quantity of antiknocking agent, tetra-ethyl lead (TEL). TEL decomposes and forms lead oxide which terminates these undesired free radical reactions. Deposition of lead can be prevented by addition of ethylene dichloride. It reacts with lead oxide to give volatile lead chloride which exit in the exhaust. Continuous exposure to lead is harmful to the health of human beings. To convert vehicular pollutants such as NO_2, CO and unburnt hydrocarbons to N_2, CO_2 and H_2O, a catalytic converter is used in cars in developed countries. The lead poisons the nobel catalyst of the converter. Thus, lead is phased out from petrol in developed countries. In India from 1995, catalytic converter is recommended in four metropolitan cities where unleaded petrol is available.

3.9.9 Unleaded Petrol

Gasoline is normally, mixed with 0.1 per cent TEL per gallon to increase octane rating by 10-15 points which have some advantages together with disadvantages. In order to overcome the harmful effects of TEL, it is slowly phased out from gasoline. The petrol free from lead is called 'unleaded petrol'. It has become mandatory to use unleaded petrol for automobiles in several parts of our country due to the following reasons:

- Lead is toxic.
- TEL is the requirement of vehicles to be fitted with catalytic converters to control pollutants. Only unleaded petrol can be used for vehicles fitted with catalytic converters because lead reduces the efficiency of the catalysts used in converters.

Unleaded petrol rating can be enhanced by the addition of certain compounds instead of TEL. Methyl-t-butyl ether (MTBE), Ethyl-t-butyl ether (ETBE), methanol, ethanol which have octane rating above 100, are added to unleaded petrol to enhance octane rating.

Cetane Number

The working of a diesel engine is different than that of a petrol engine. In diesel engine fuel is ignited by application of heat and pressure and not by spark.

Two hydrocarbons α-methyl naphthalene and n-cetane/n-hexadecane are specified as standards. Since n-cetane has low ignition lag, its cetane number is arbitrarily fixed as 100, while α-methyl naphthalene has long ignition lag and its cetane number is fixed as zero.

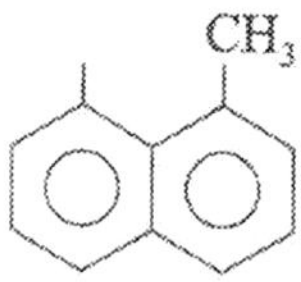

$$CH_3-(-CH_2-)_{14}-CH_3$$

α-Methylnaphthalene
Caetane number 0

n-cetane
Caetane number 100

The cetane number of a diesel fuel is defined as the percentage of n-cetane in the mixture of n-cetane and α-methyl naphthalene. Higher the cetane number, more is the ignitability.

Thus, the structural requirements for a diesel fuel are precisely those which have to be avoided in gasoline.

- The straight chain alkanes like n-cetane, which ignite readily are good diesel fuels. Aromatics like α-methyl naphthalene which has long ignition delay is a poor diesel fuel. Cetane numbers in practice range from about 25-50, increasing with increasing engine speed.
- Certain additives can be used to raise the cetane ratings. Isoamyl nitrates such as ethyl nitrate ($CH_3 - CH_2 - O - NO_2$) and acetone are used to increase cetane number. The fuel has cetane number 70 when it matches with ignition properties of the mixture of 70% cetane and 30% α-methyl naphthalene.

3.10 POWER ALCOHOL

When ethyl alcohol is used as an additive to motor fuels in internal combustion engines, it is called power alcohol. Generally, 5-25% alcohol is mixed with petrol.

Manufacture of Power Alcohol

The raw materials for the manufacture of power alcohol are molasses, sugar beets, sugar cane, etc. are saccharine materials, starchy materials such as starch, potatoes, cereal grains, etc. cellulose materials such as sulphite liquor from paper mills are used with hydrocarbon gases.

Ethyl alcohol from molassses

Dark colored viscous liquid left after the crystallization of cane sugar from cane juice contains 50 55% total sugar, of which 35-40% is cane sugar or sucrose is called molasses. It is converted into ethyl alcohol by means of yeast, which contains enzymes invertase and zymase responsible for fermentation.

$$\underset{\text{Sucrose in molasses}}{C_{12}H_{22}O_{11}} \xrightarrow{\text{Invertase}} \underset{\text{Glucose}}{C_6H_{12}O_6} + \underset{\text{Fructose}}{C_6H_{12}O_6}$$

$$\underset{\text{Glucose, Fructose}}{C_6H_{12}O_6} \xrightarrow{\text{Zymase}} 2C_2H_5OH + 2CO_2$$

The molasses is diluted with water to bring down the concentration of sugar to 10-12%. pH is kept between 4 and 5 to favour the function of enzymes, by using sulphuric acid. After fermentation with yeast for 48-60 hours, it is distilled to get ethyl alcohol. Absolute alcohol can be prepared by repeated distillation.

3.11 BIODIESEL

Vegetable oils and animal fats are basically triglycerides. They are fatty acid esters of glycerol. Vegetable and animal fats have very high viscosity and they cannot be used directly as a fuel. Triglycerides can be converted into their methyl or ethyl esters by transesterification.

Moisture free oil or fat (with less than 5% free acid) and dry methanol or ethanol are taken in 1:3:5 molar ratio in a three-neck round bottom flask (with arrangement of magnetic stirrer and heating) and catalyst sodium methoxide maintained about 2% by weight. The mixture is heated at 60°C, and purified by using vacuum distillation.

$$\underset{\text{Veg. Oil or fat}}{\begin{array}{l} CH_2-OOC-R_1 \\ | \\ CH-OOC-R_2 \\ | \\ CH_2-OOC-R_3 \end{array}} \xrightarrow[\text{Heat}\ 60°C]{CH_3ONa} \underset{\text{Biodiesel}}{\begin{array}{l} R_1-COO\,CH_3 \\ \\ R_2-COO\,CH_3 \\ \\ R_3-COO\,CH_3 \end{array}} + \underset{\text{Glycerol}}{\begin{array}{l} CH_2OH \\ | \\ CHOH \\ | \\ CH_2OH \end{array}}$$

Advantages

1. can be used as a fuel cetane number 46 to 54 and high calorific value about 40 kJ/gm
2. less pollution
3. clean to use in diesel engines
4. is cheaper

3.12 POWER ALCOHOL AS A FUEL

In order to obtain better results, power alcohol is used as a fuel by blending (up to 25%) with petrol in internal combustion engines as a motor spirit. Industrial alcohol containing 95% alcohol and 5% water can also be blended with petrol by using blending agents like benzene, ether, etc. Without the use of blending agent application is difficult as industrial alcohol does not mix with petrol.

Advantages of alcohol-blended petrol

- Better antiknock properties because octane number of alcohol is 90; while petrol has octane number of 60-70. Addition of power alcohol to petrol increases the octane number. Alcohol-blended petrol can be used in engines with higher compression ratio because of the higher octane number. This compensates for the lower heating value of alcohol in the blend.
- There are no starting difficulties with alcohol-petrol blend.
- Lubrication in case of pure petrol and alcohol petrol blend is the same.
- Amount of air required for complete combustion is less.

Disadvantages of alcohol-blended petrol

- Calorific value of petrol is lowered because of alcohol.
- Specific arrangements for starting alcohol-petrol-blend in carburettor is necessary because of considerable surface tension of alcohol, which creates difficulties in atomisation of alcohol at low temperature.
- Alcohol may cause corrosion at it is easily oxidised to acids.
- As less air is required for combustion, modification in air is needed.
- Separation of alcohol and petrol layers takes place especially at low temperature as alcohol absorbs moisture. Blending agents such as benzene or toluene are used to avoid this.

SOLVED PROBLEMS

Problem 1: A sample of coal on analysis was found to contain C = 60%, O = 33%, H = 6%, N = 0.2%, S = 0.5%, Ash = 0.3%. Calculate the higher and lower calorific value of the coal by using Dulong's formula.

Solution:

$$\text{H.C.V.} = \frac{1}{100}\left[8080\,\text{C} + 34{,}500\left(\text{H} - \frac{\text{O}}{8}\right) + 2240\,\text{S}\right] \text{cal/gm}$$

$$= \frac{1}{100}\left[8080 \times 60 + 34{,}500\left(6 - \frac{33}{8}\right) + 2240 \times 0.5\right]$$

$$= \frac{1}{100}[484800 + 64687.5 + 1120]$$

$$= 5506 \text{ cal/gm}$$

$$\text{L.C.V.} = \text{H.C.V.} - 0.09 \text{ H} \times 587$$

$$= 5506 - 0.09 \times 6 \times 587 = 5506 - 316.98$$

$$= 5189.02 \text{ cal/gm}$$

Problem 2: A sample of coal was found to contain following by mass C = 80%, N = 1.5%, H = 6%, S = 1.5%, O = 8%, ash = 3.0%. Calculate Gross and Net calorific value of coal by using Dulongs formula.

Solution:

$$\text{G.C.V.} = \frac{1}{100}\left[8080 \text{ C} + 34{,}500\left(\text{H} - \frac{\text{O}}{8}\right) + 2240 \text{ S}\right] \text{ cal/gm}$$

$$= \frac{1}{100}\left[8080 \times 80 + 34{,}500\left(6 - \frac{8}{8}\right) + 2240 \times 1.5\right]$$

$$= \frac{1}{100}[646400 + 207000 + 3360]$$

$$= 8567.60 \text{ cal/gm}$$

$$\text{N.C.V.} = [\text{H.C.V.} - 0.09 \text{ H} \times 587] \text{ cal/gm}$$

$$= 8567.6 - 0.09 \times 6 \times 587$$

$$= 8567.6 - 316.98 \text{ cal}$$

$$= 8250.6 \text{ cal/gm}$$

Problem 3: A sample of coal contains C = 80%, H = 7%, O = 3%, S = 3.5%, N = 2.1% and ash = 4.4%. Calculate the gross and net calorific value of coal. (Dec. 2005)

Solution:

$$\text{H.C.V.} = \frac{1}{100}\left[8080 \text{ C} + 34{,}500\left(\text{H} - \frac{\text{O}}{8}\right) + 2240 \text{ S}\right]$$

$$= \frac{1}{100}\left[8080 \times 80 + 34{,}500\left(7 - \frac{3}{8}\right) + 2240 \times 3.5\right]$$

$$= \frac{1}{100}[646400 + 34500\,(7 - 0.375) + 2240 \times 3.5]$$

$$= \frac{1}{100}[646400 + 228562.5 + 7840]$$

$$= \frac{1}{100}[8828025]$$

$$= 8828.025 \text{ Kcal/kg}$$

$$\text{L.C.V.} = \text{H.C.V.} - 0.09\ \text{H} \times 587 \text{ kcal/kg}$$

$$= 8828.025 - 0.09 \times 7 \times 587$$

$$= 8828.025 - 369.81 = 8458.21 \text{ kcal/kg}$$

Problem 4: A sample of coal contains C = 61%, O = 32%, H = 6%, S = 0.5%, N = 0.2%, Ash = 0.3%. Calculate H.C.V. and L.C.V using Dulong's formula. (Dec. 2004)

Solution:

$$\text{H.C.V.} = \frac{1}{100}\left[8080\,\text{C} + 34{,}500\left(\text{H} - \frac{\text{O}}{8}\right) + 2240\,\text{S}\right]$$

$$= \frac{1}{100}\left[8080 \times 61 + 34{,}500\left(6 - \frac{32}{8}\right) + 2240 \times 0.5\right]$$

$$= \frac{1}{100}[492880 + 69000 + 1120]$$

$$= 5630 \text{ kcal/kg}$$

$$\text{L.C.V.} = \text{H.C.V.} - 0.09\ \text{H} \times 587 \text{ kcal/kg}$$

$$= 5630 - 0.09 \times 6 \times 587$$

$$= 5630 - 316.98$$

$$= 5313.02 \text{ kcal/kg}$$

Problem 5: Calculate the volume of oxygen required for complete combustion of 2 m^3 of a gaseous fuel of following composition by volume CO = 23%, H_2 = 12%, N_2 = 50%, CO_2 = 5%, O_2 = 2%, CH_4 = 3%.

Solution: 1m^3 of fuel contains

$CO = 23\% = 0.23\ m^3$

$H_2 = 12\% = 0.12\ m^3$

$N_2 = 50\% = 0.50\ m^3$

$CO_2 = 5\% = 0.05\ m^3$

$O_2 = 2\% = 0.02\ m^3$

$CH_4 = 3\% = 0.03\ m^3$

$$\underset{1\text{ vol}}{CO} + \underset{0.5\text{ vol}}{1/2\,O_2} \longrightarrow \underset{1\text{ vol}}{CO_2}$$

$$\underset{1\text{ vol}}{H_2} + \underset{0.5\text{ vol}}{1/2\,O_2} \longrightarrow \underset{1\text{ vol}}{H_2O}$$

$$\underset{1\text{ vol}}{CH_4} + \underset{2\text{ vol}}{2O_2} \longrightarrow \underset{1\text{ vol}}{CO_2 + 2H_2O}$$

O_2 required for 0.23 vol. of CO $= 0.5 \times 0.23 = 0.115\ m^3$

O_2 required for 0.12 vol. of H_2 $= 0.5 \times 0.12 = 0.06\ m^3$

O_2 required for 0.03 vol. of CH_4 $= 2 \times 0.03 = 0.06\ m^3$

Total O_2 required $= 0.235\ m^3$

Net volume of O_2 required = Theoretical vol. of O_2 required – O_2 present in the fuel.

$$= 0.235\ m^3 - 0.02\ m^3 = 0.215\ m^3$$

Amount of O_2 required for 2 m^3 of gaseous fuel

$$= 0.215\ m^3 \times 2 = 0.430\ m^3.$$

Problem 6: Calculate the amount of air required per kilogram for complete combustion of oil. The oil contains 85% carbon and 15% hydrogen. Also calculate percentage of amount of CO_2 produced, assuming the air contains 20% oxygen and 80% nitrogen by mass.

Solution: 1 kg of oil contains 0.85 kg carbon
0.15 kg hydrogen

The combustion equations are

1. $\underset{12}{C} + \underset{32}{O_2} \longrightarrow \underset{44}{CO_2}$

2. $\underset{2}{H_2} + \underset{16}{\frac{1}{2}O_2} \longrightarrow \underset{18}{H_2O}$

∴ the minimum amount of oxygen required for combustion

$$= (\text{Mass of carbon}) \times \frac{32}{12} + (\text{Mass of hydrogen}) \times \frac{16}{2} - (\text{Mass of oxygen present})$$

$$= 0.85 \times \frac{32}{12} + 0.15 \times \frac{16}{2} - 0$$

$$= 2.266 + 1.2$$

$$= 3.466\ kg$$

It is given that air contains 20% by mass of oxygen.

∴ the amount of air required for complete combustion

$$= 3.466 \times \frac{100}{20} = 17.330\ kg$$

Percentage composition of products of combustion by mass

Products of combustion: CO_2

[a] CO_2 = Mass of C $\times \frac{44}{12}$

$$= 0.85 \times \frac{44}{12}$$

$$= 3.116 \text{ kg (Amount of } CO_2 \text{ produced)}$$

Problem 7: The percentage composition of a coal sample is as follows: C = 90.0%, H = 3.5%, O = 3.0%, N = 1.0%, S = 0.5%. Calculate the theoretical volume of oxygen and air required at 27°C and 1 atmospheric pressure for combustion of 1 kg of fuel.

Solution: 1 kg of coal will contain 0.90 kg of carbon
0.035 kg of hydrogen
0.030 kg oxygen
0.005 kg sulphur

Equations for calculations are

1. $$\underset{12}{C} + \underset{32}{O_2} \longrightarrow \underset{44}{CO_2}$$

2. $$\underset{2}{H_2} + \underset{16}{\frac{1}{2}O_2} \longrightarrow \underset{18}{H_2O}$$

Oxygen required

$$= \frac{32}{12} \times \text{Carbon} + \frac{16}{2} \times \text{Hydrogen} + \frac{32}{32} \times \text{Sulphur} - (\text{oxygen present in fuel})$$

$$= \frac{32}{12} \times 0.90 + \frac{16}{2} \times 0.35 + \frac{32}{32} \times 0.005 - 0.03$$

$$= 2.685 - 0.03$$

$$= 2.655 \text{ kg of oxygen}$$

Oxygen required for complete combustion of 1 kg of fuel

$$= 2.655 \text{ kg at STP}$$

To find the volume at 27°C and 1 atmospheric pressure we use the following equations

$$\frac{P_1 V_1}{V_1} = \frac{P_2 V_2}{V_2}$$

and we known that

32 gm of oxygen occupies 22.4 litres at STP

$\therefore$ 2.655 gm of oxygen occupies $\dfrac{22.4 \times 2.655}{32}$

$$= 1858.5 \text{ litre}$$

$P_1 = 760$ mm $\quad P_2 = 760$ mm

$T_1 = 273°K \quad T_2 = 27 + 273 = 300°C$

$V_1 = 1858$ litres $\quad V_2 = ?$

$$\therefore \quad V_2 = \frac{P_1 V_1}{V_1} \times \frac{T_2}{P_2}$$

$$= \frac{V_1}{T_1} \times T_2 \text{ Since } P_1 = P_2$$

$$V_2 = \frac{V_1 T_2}{T_1}$$

$$= \frac{1858.5 \times 300}{273}$$

$$= 2042.30 \text{ litres}$$

∴ The volume of oxygen required for complete combustion of 1 kg of coal at 27°C and 1 atmospheric pressure is 2042 litres.

Volume of oxygen required at STP = 1858.5 litres

Volume of air at STP

$$= 1858.5 \times \frac{100}{23}$$

$$= 8080.43 \text{ litres at STP}$$

Volume at 27°C and 1 atmospheric pressure

$P_1 = 760$ mm $\quad P_2 = 760$ mm

$V_1 = 8080.43 \quad V_2 = ?$

$T_1 = 273°K \quad T_2 = 27 + 273 = 300°C$

$$\therefore \quad \frac{P_1 V_1}{V_1} = \frac{P_2 V_2}{V_2}$$

$$V_2 = \frac{P_1 V_1}{V_1} \times \frac{T_2}{P_2}$$

$$= \frac{V_1}{T_1} \times T_2$$

$$V_2 = \frac{8080.73 \times 300}{273}$$

$$= 8879.59 \text{ litres}$$

Problem 8: The percentage composition by mass of a sample of coal is as follows: C = 81%, H = 5%, O = 8.5%, S = 1%, N = 1%, ash = rest.

Calculate the (a) amount of air required for complete combustion of 1 kg of fuel and (b) gross and net calorific value of fuel.

Given calorific value of C = 8960 Kcal/Kg

H = 34000 Kcal/Kg

S = 2200 Kcal/Kg

and latent heat of steam = 540 Kcal/Kg

Solution: Combustion equations

1. $\underset{12}{C} + \underset{32}{O_2} \longrightarrow \underset{44}{CO_2}$

2. $\underset{2}{H_2} + \underset{16}{\frac{1}{2}O_2} \longrightarrow \underset{18}{H_2O}$

3. $\underset{32}{S} + \underset{32}{O_2} \longrightarrow \underset{64}{SO_2}$

1 kg of fuel contains 0.81 kg carbon
0.05 kg hydrogen
0.085 kg oxygen
0.01 kg sulphur

Amount of oxygen required

$$0.81 \times \frac{32}{12} = 2.16 \text{ kg}$$

$$0.05 \times \frac{16}{2} = 0.40 \text{ kg}$$

$$0.01 \times \frac{32}{32} = 0.01 \text{ kg}$$

$$= 2.570 \text{ kg}$$

Oxygen present = 0.085 kg

Oxygen required = 2.570 kg – 0.085 kg
= 2.485 kg

∴ Amount of air required

$$= 2.485 \times \frac{100}{23}$$

$$= 10.80 \text{ kg}$$

To find out the gross and net calorific value we use Dulong's formula

$$\text{G.C.V.} = \left[8960 \times 0.81 + 34000\left(H - \frac{O}{8}\right) + 0.01 \times 2200\right]$$

$$= \left[7257.6 + 34000\left(0.05 - \frac{0.085}{8}\right) + 22\right]$$

$$= 7257.6 + 1339.6 + 22.0 = 8619.2 \text{ kcal/kg}$$

Net calorific value

$$= \text{Gross calorific value} - 0.09\,H \times 540$$

$$= 8619.2 - 0.09 \times 5 \times 540 = 8619.2 - 243.0 = 8376.2 \text{ kcal/kg}$$

Results

1. Amount of air required for complete combustion = 10.80 kg
2. Gross calorific value of fuel = 8619.2 kcal/kg
3. Net calorific value = 8376 kcal/kg

Problem 9: A gaseous fuel has the following composition by volume $H_2 = 40\%$, $CH_4 = 40\%$, $C_2H_6 = 6\%$, $CO = 6\%$, $N_2 = 1\%$. Calculate the volume of air required for complete combustion of $1 m^3$ of fuel.

Solution: $1 m^3$ of fuel will contain

$0.400\ m^3$ of hydrogen

$0.400\ m^3$ of CH_4

$0.060\ m^3$ of C_2H_6

$0.060\ m^3$ of CO

$0.01\ m^3$ of N_2

Combustion equations are

1. $H_2 + \frac{1}{2}O_2 \longrightarrow H_2O$

   1 vol $\quad \frac{1}{2}$ vol $\quad$ 1 vol

2. $CH_4 + 2O_2 \longrightarrow CO_2 + 2H_2O$

   1 vol $\quad$ 2 vol $\quad$ 1 vol $\quad$ 2 vol

3. $C_2H_6 + \frac{7}{2}O_2 \longrightarrow 2CO_2 + 3H_2O$

   1 vol $\quad \frac{7}{2}$ vol $\quad$ 2 vol $\quad$ 3 vol

4. $CO + \frac{1}{2}O_2 \longrightarrow CO_2$

1 vol $\quad \frac{1}{2}$ vol $\quad$ 1 vol

∴ Volume of oxygen required

$$= 0.400 \times \left(\frac{1}{2}\right) + 0.40\ (2) + 0.06\left(\frac{7}{2}\right) + 0.06\left(\frac{1}{2}\right)$$
$$= 0.20 + 0.80 + 0.21 + 0.03$$
$$= 1.24 \text{ m}^3 \text{ of oxygen}$$

Air contains 21% by volume of oxygen.

∴ 1.24 m^3 of oxygen corresponds to

$$= 1.24 \times \frac{100}{21}$$
$$= 5.906 \text{ m}^3.$$

Problem 10: On analysis a flue gas contained CO_2 = 10.75%, CO = 1.16%, O_2 = 6.8%, N_2 = 81.39% Chemical analysis of coal gas C = 88, H = 3.5%, O = 4.5%, ash = 4%. Determine how much of air was supplied to burn 1 kg of fuel?

Solution: 1 kg of coal contains 0.88 kg of carbon
0.035 kg of hydrogen
0.045 kg of oxygen
0.040 kg of ash

Combustion equations are

1. $C + O_2 \longrightarrow CO_2$
$\quad 12 \quad 32 \quad\quad 44$

2. $H_2 + \frac{1}{2}O_2 \longrightarrow H_2O$
$\quad 2 \quad 16 \quad\quad 18$

∴ Amount of oxygen required

$$= 0.88 \times \frac{32}{12} + 0.035 \times \frac{16}{2} - 0.045$$
$$= 2.346 + 0.28 - 0.045$$
$$= 2.581 \text{ kg}$$

The % by weight of air is 23

$\therefore$ 2.581 kg of oxygen corresponds

$$2.581 \times \frac{100}{23} = 11.22 \text{ kg of air}$$

In the flue gas analysis there is 6.8% oxygen. This means excess air is used for combustion. Assuming the % composition is on mass basis.

The amount of CO_2 according to carbon content is

$$0.88 \times \frac{44}{12} = 3.226 \text{ kg}$$

% CO_2 in the flue gas is 10.75%, equating these two values we get the total amount of flue gas

$$\frac{3.226}{x} \times 100 = 10.75$$

$\therefore$ The amount of oxygen in the flue gas can now be determined

$$\frac{y}{30.01} \times 100 = 6.8$$

$$\therefore \qquad y = \frac{6.8 \times 30.01}{100} = 2.04 \text{ kg}$$

This oxygen present should be excess oxygen supplied in air for combustion.

The amount of oxygen theoretically required

For combustion	2.581 kg
Excess oxygen	2.040 kg
Total oxygen supplied	4.621 kg (2.581 + 2.04)

$\therefore$ Total air supplied $\qquad 4.621 \times \frac{100}{23} = 20.0869 \text{ kg}$

Result: Mass of air for complete combustion of fuel = 20.0869 kg

Problem 11: A fuel has the following composition by volume. Hydrogen 24%, 11%, ethane, methane 30%, ethene 4.5%, butene 2.5%, carbon monoxide 6.0%, carbon dioxide 8%, oxygen 2% and nitrogen 12%. Calculate the theoretical amount of air required at 25°C and 750 mm pressure for complete combustion of 1m^3 of fuel.

Solution: 1m^3 of fuel gas contain $\quad$ 0.24 m^3 hydrogen

0.30 m^3 methane

0.11 m^3 ethane

0.045 m^3 ethene

0.025 m^3 butene

0.06 m^3 carbon monoxide

$0.08\ m^3$ carbon dioxide
$0.02\ m^3$ oxygen
$0.12\ m^3$ nitrogen

Combustion equations are

1. $H_2(g) + \frac{1}{2}O_2(g) \longrightarrow H_2O(l)$

   1 vol $\quad \frac{1}{2}$ vol $\quad$ 1 vol

2. $CH_4(g) + 2O_2(g) \longrightarrow CO_2 + 2H_2O(l)$

   1 vol $\quad$ 2 vol $\quad$ 1 vol

3. $C_2H_6(g) + \frac{7}{2}O_2(g) \longrightarrow 2CO_2(g) + 3H_2O(l)$

   1 vol $\quad \frac{7}{2}$ vol $\quad$ 2 vol

4. $C_2H_4(g) + 3O_2(g) \longrightarrow 2CO_2(g) + 2H_2O(l)$

   1 vol $\quad$ 3 vol $\quad$ 2 vol

5. $C_4H_4(g) + 6O_2(g) \longrightarrow 4CO_2(g) + 4H_2O(l)$

   1 vol $\quad$ 6 vol $\quad$ 4 vol

6. $CO(g) + \frac{1}{2}O_2(g) \longrightarrow CO_2(g)$

   1 vol $\quad$ 0.5 vol $\quad$ 1 vol

∴ Volume of oxygen required for combustion

$$= 0.24 \times \left(\frac{1}{2}\right) + 0.3 \times 2 + 0.11 \times 3.5 + 0.045 \times 3 + 0.025 \times 6 + 0.06 \times 0.05 - 0.02\ m^3$$

$$= 0.12 + 0.600 + 0.385 + 0.135 + 0.150 + 0.03 - 0.02\ m^3$$
$$= 1.4\ m^3$$

Corresponding oxygen at 25°C and 750 mm pressure can be calculated by using the equation

$$\frac{P_1V_1}{T_1} = \frac{P_2V_2}{T_2}$$

$P_1 = 760, \quad P_2 = 750$
$T_1 = 273\ °K, T_2 = 25 + 273 = 298°C$
$V_1 = 1.4\ m^3, V_2 = ?$

$$\frac{760 \times 1.4}{273} = \frac{750 \times V_2}{298}$$

$$V_2 = \frac{760 \times 1.4 \times 298}{273 \times 750}$$

$$= 1.548 \text{ m}^3$$

Volume of air required $= 1.548 \times \frac{100}{21} = 7.37 \text{ m}^3$.

Problem 12: A coal sample was found to contain the following, C: 80 per cent, H_2: 4 per cent, O_2: 2 per cent, N_2: 1 per cent and the remaining being ash. Estimate the quantity of minimum air required for the complete combustion of 1 kg of coal sample if 20 per cent of excess air is supplied.

Solution: 1 kg of coal contains 0.80 kg of C, 0.041 kg of H_2, 0.021 kg of O_2 and 0.01 kg of N_2

$$\underset{12}{C} + \underset{32}{O_2} \longrightarrow \underset{44}{CO_2}$$

$$\underset{2}{H_2} + \underset{16}{O} \longrightarrow \underset{18}{H_2O}$$

12 kg of C requires 32 kg of O_2.

0.80 kg of C requires $= \frac{32}{12} \times 0.80 = 2.133$ kg of O_2.

2 kg of H_2 requires 16 kg of O_2.

0.041 kg of H_2 requires $= \frac{16}{2} \times 0.041 = 0.328$ kg of O_2.

N_2 does not burn. Hence, it does not consume any O_2. Theoretical O_2 required for

1 kg of the fuel $= 2.133 + 0.328 = 2.461$ kg of O_2

Net amount of O_2 required for 1 kg of the fuel = Theoretical O_2 required – O_2 present in the fuel

$= 2.461 - 0.02 = 2.441$ kg of O_2

23 kg of O_2 is supplied by 100 kg of air

3.441 kg of O_2 is supplied by $= \frac{100}{23} \times 2.441 = 10.613$ kg of air

If 20 per cent excess air is used, amount of air required for the complete combustion of 1 kg of coal

$$= \frac{10.613}{100} \times 120 = 12.735 \text{ kg of air}$$

Problem 13: Calculate the minimum weight of air required for the complete combustion of 1 kg of coal having the following percentage composition by weight C: 90.1%, H_2: 3.4%, O_2: 3%, S: 0.5%, N_2: 0.5% and ash 1.5% Calculate the percentage composition by weight of the dry products of combustion.

Solution: 1 kg of coal contain 0.91 kg of C, 0.034 kg of H_2, 0.005 kg of S, 0.03 kg of O_2 and 0.005 kg of N_2.

$$\underset{12}{C} + \underset{32}{O_2} \longrightarrow \underset{44}{CO_2}$$

$$\underset{2}{H_2} + \underset{16}{O} \longrightarrow \underset{18}{H_2O}$$

$$\underset{32}{S} + \underset{32}{O_2} \longrightarrow \underset{64}{SO_2}$$

12 kg of C requires 32 kg of O_2.

$$0.91 \text{ kg of C requires} = \frac{32}{12} \times 0.91$$

$$= 2.426 \text{ kg of } O_2$$

2 kg of H_2 requires 16 kg of O_2.

$$0.034 \text{ kg of } H_2 \text{ requires} = \frac{16}{2} \times 0.034 = 0.272 \text{ kg of } O_2.$$

32 kg of S requires 32 kg of O_2.

$$0.005 \text{ kg of S requires} = \frac{32}{32} \times 0.005 = 0.005 \text{ kg of } O_2.$$

N_2 does not burn. Hence, it does not consume any O_2.

Theoretical required for 1 kg of the fuel – 2.426 + 0.272 + 0.005

$$= 2.7036 \text{ kg of } O_2$$

Net amount of O_2 required for 1 kg of the fuel = Theoretical O_2 required – O_2 present in the fuel

$$= 2.7036 - 0.03 = 2.6736 \text{ kg of } O_2$$

23 kg of O_2 is supplied by 100 kg of air.

$$2.6736 \text{ kg of } O_2 \text{ is supplied by} = \frac{100}{23} \times 2.6736 = 11.62 \text{ kg}$$

Masses of the dry products of combustion

12 kg of C gives 44 kg of CO_2

$$0.91 \text{ kg of C gives} = \frac{44}{12} \times 0.91 = 3.336 \text{ kg of } CO_2.$$

32 kg of S gives 64 kg of SO_2

0.005 kg of S gives $= \frac{64}{32} \times 0.005$

$= 0.01$ kg of SO_2.

Mass of N_2 = N_2 present in the fuel + N_2 present in the air

$$= 0.005 + 11.54 \times \frac{77}{100}$$

$$= 0.005 + 8.886$$

$= 8.891$ kg of N_2

Total mass of combustion products = 3.33 + 0.01 + 8.891 = 12.23 kg

Percentage composition of the dry products of combustion:

$$\% \text{ of } CO_3 = \frac{3.3}{12.23} \times 100$$

$$= 26.98\%$$

$$\% \text{ of } SO_2 = \frac{0.01}{12.23} \times 100$$

$$= 0.0817\%$$

$$\% \text{ of } N_2 = \frac{8.891}{12.23} \times 100$$

$$= 72.69\%$$

Problem 14: A coal sample on analysis gives the following, C: 66.0%, H_2: 4.0%, S: 2.5%, N_2: 1.4%, O_2: 6.1% and ash 9.5%. Calculate the percentage composition of the dry products of combustion if 1 kg of coal is burnt with 25% excess of air.

Solution: 1 kg of coal contains 0.662 kg of C, 0.04 kg of H_2, 0.025 kg of S, 0.061 kg of O_2 and 0.014 kg of N_2.

$$\underset{12}{C} + \underset{32}{O_2} \longrightarrow \underset{44}{CO_2}$$

$$\underset{2}{H_2} + \underset{16}{O} \longrightarrow \underset{18}{H_2O}$$

$$\underset{32}{S} + \underset{32}{O_2} \longrightarrow \underset{64}{SO_2}$$

12 kg of C requires 32 kg of O_2

0.660 kg of C requires $= \frac{32}{12} \times 0.660 = 1.7599$ kg of O_2

2 kg of H_2 requires 16 kg of O_2

$$0.040 \text{ kg of } H_2 \text{ requires} = \frac{16}{2} \times 0.040 = 0.32 \text{ kg of } O_2$$

32 kg of S requires 32 kg of O_2

$$0.025 \text{ kg of S requires} = \frac{32}{32} \times 0.025 = 0.025 \text{ kg of } O_2.$$

N_2 does not burn. Hence it does not consume any O_2,

Theoretical O_2 required for

1 kg of the fuel = 1.7599 + 0.32 + .025 = 2.1049 kg of O_2

Net amount of O_2 required for 1 kg of the fuel = Theoretical O_2 required – O_2 present in the fuel.

$$= 2.1049 - 0.061 = 2.0439 \text{ kg of } O_2.$$

23 kg of O_2 is supplied by 100 kg of air

$$2.0439 \text{ kg of } O_2 \text{ is supplied by} = \frac{100}{23} \times 2.0439 = 8.8865 \text{ kg}$$

If 25% excess air is used, amount of air required for the complete combustion of 1 kg of coal $= \frac{8.8865}{100} \times 125 = 11.108$ kg.

Masses of the dry products of combustion

12 kg of C gives 44 kg of CO_2.

$$0.660 \text{ kg of C gives} = \frac{44}{12} \times 0.660$$

$$= 2.4199 \text{ kg of } CO_2$$

32 kg of S gives 64 kg of SO_2

$$0.025 \text{ kg of S gives} = \frac{64}{32} \times 0.025$$

$$= 0.050 \text{ Kg of } SO_2$$

Mass of N_2 = N_2 present in the fuel + N_2 present in the air

$$= 0.014 + \frac{77}{100} \times 11.25$$

$$= 0.014 + 8.66$$

$$= 8.674 \text{ Kg of } N_2$$

$$\text{Mass of } O_2 = (11.25 - 9.0) \times \frac{23}{100} = 0.5 \text{ kg}$$

Total mass of combustion products $= 2.42 + 0.050 + 8.674 + 0.5$

$= 11.644$ kg

Percentage composition of the dry products of combustion:

$$\% \text{ of } CO_2 = \frac{2.4199}{11.644} \times 100$$
$$= 20.78\%$$

$$\% \text{ of } SO_2 = \frac{0.050}{11.644} \times 100$$
$$= 0.4294\%$$

$$\% \text{ of } O_2 = \frac{0.5}{11.644} \times 100$$
$$= 4.294\%$$

$$\% \text{ of } N_2 = \frac{8.674}{11.644} \times 100$$
$$= 74.49\%$$

QUESTIONS

- How are fuels classified? What is meant by gross calorific value of a fuel?
- Compare the advantages of gaseous fuels over liquid fuels.
- How is coal analysed for evaluating the carbon, hydrogen and sulphur content?
- Define Octane number and Cetane number.
- What is proximate analysis of coal? How is it carried out? What is the significance of the analysis?
- What is meant by 'calorific value'?
- Define net and gross calorific value of a fuel? How are these values determined for a solid fuel?
- What is petroleum? How is it classified?
- How is crude petroleum oil refined? Discuss the distillation process and the products obtained thereof.
- What is meant by cracking? Discuss any one method of cracking.
- What is thermal cracking? How does it differ from catalytic cracking?
- Explain the term 'cracking'. What is catalytic cracking? Explain the advantages of catalytic cracking over thermal cracking.
- How many methods are used in catalytic cracking? Discuss any one method in detail.
- How ultimate analysis of coal is carried out? What is its significance?
- Give synthesis of (i) Alcohol from molasses (ii) Biodiesel.

ADDITIONAL PROBLEMS FOR PRACTICE

- Calculate gross and net calorific value of the fuel which on ultimate analysis gave the following result C = 90%, H = 5%, O = 3%, N = 1%, S = 0.3% and ash = 0.7%.
- A sample of coal was found to contain C = 76%, O = 16%, N = 1.5%, S = 0.3%, H = 5% and ash = 1.0%. Calculate higher and lower calorific value assuming latent heat of condensation of steam 580 kcal/kg.
- A coal sample has the following composition C = 84.1%, H = 3.4%, N = 0.6%, S = 0.4%, O = 3.1%, moisture = 3.5% and remaining ash. Calculate: (i) Amount of air required for complete combustion of 1 kg of coal. (ii) Percentage composition by weight of the dry products of combustion when 1 kg of coal has been burnt completely.
- A composition of a coal sample is C = 80.5%, H = 5.5%, O = 8.0%, S = 1.5%, N = 1.5% and remaining ash. Calculate (i) H.C.V. and L.C.V (ii) Amount of air required for complete combustion of 1 kg of coal. Calorific values of C = 8060 kcal/kg, H = 34000 kcal/kg and S = 2200 kcal/kg
- A coal sample has the following composition C = 52.5%, O = 8.1%, H_2 = 25%, S = 1.0%, N = 3.9% and remaining nitrogen. Calculate (i) H.C.V. and L.C.V. (ii) Amount of air required for complete combustion of 2.5 kg of coal. (iii) Weights of combustion products.
- A coal sample with composition C = 63.5%, H = 5.7%, O = 9.2%, N = 0.6%, S = 0.8%, moisture = 15.1% and remaining ash. Calculate (i) H.C.V. and L.C.V. (ii) weight of combustion products when 1 kg of coal has been burnt completely and amount of air required.
- Coal sample on proximate analysis gave the following results: 99 gm of air dired coal sample was taken in silica crucible and heated in an electric oven at 105-110°C for 1 hr, residue was weighed 4.821 gm. The residue was then ignited at 725-750°C to a constant weight of 0.4938 g. In another experiment 1.05 g of sample was heated in silica crucible covered with lid at 950°C for exactly 7 mins. After cooling weight of residue was found to be 0.641 g. Calculate % of fixed carbon.
- On ultimate analysis

  (i) 0.125 g of sample on combustion gave 0.397 g CO_2 and 0.0109 g H_2O.

  (ii) 2.81 gm of sample in estimation of nitrogen liberated NH_3 was absorbed in 50 ml of 0.1 N H_2SO_4. The resultant solution required 20.1 ml 0.1 N NaOH for complete neutralisation. The blank titration reading was 48.9 ml.

  (iii) 6.40 g of sample in determination of sulphur gave 0.466 g $BaSO_4$.

  Calculate (i) G.C.V. and N.C.V. (ii) Amount of O_2 and air required for complete combustion of 1 kg of coal. (iii) Also calculate volume of O_2 and air at 750 mm pressure and 25°C temperature.

MUMBAI UNIVERSITY QUESTIONS

May 2002

- 2.4995 g of air dried coal sample was taken in a silica crucible, after heating it in an electric oven at 105-110°C for 1 hr, the residue was weighed 2.4147 g. Calculate the percentage of moisture in coal.
- Calculate the mass of oxygen required per kilogram of oil containing 84.5% carbon and 15.5% hydrogen. Also calculate weights of the contribution products CO_2 and H_2O.
- What is cracking? Explain catalytic cracking.

Dec. 2002

- What is Dulong's formula?
- The composition of a gas was found to be H_2 = 10%, CH_4 = 16%, C_2H_6 = 20%, N_2 = 6%, CO = 22%, CO_2 = 18% and remaining is oxygen by volume. Calculate the volume of oxygen and air required for complete combustion of 10 m^3 of this gas.
- Write any six differences between catalytic and thermal cracking.

May 2003

- What is knocking? What are antiknocking agents?
- By Kjeldahl's method 1.5 gm of a sample of coal was analyzed. The ammonia evolved was absorbed in 50 ml of 0.1 N H_2SO_4. After absorption the excess H_2SO_4 required 35 ml of 0.1 N NaOH for neutralisation. Calculate the percentage nitrogen.
- A sample of coal contains C = 70%, O = 20%, H = 8%, S = 1%, N = 0.5% and ash = 0.5%. Calculate GCV and NCV of the coal.
- Write short notes on octane and cetane number.
- A sample of coal contains C = 70%, H = 10%, O = 05%, S = 3%, Ash = remaining and N = 2%. Calculate the minimum amount of air needed for complete combustion of 1 kg of the coal.

Dec. 2003

- What are the characteristics of good fuel?
- Differentiate between octane number and cetane number.
- The % composition by mass of a sample of coal is as follows: C = 80%, H = 6%, O = 8%, S = 1.5%, N = 1%, ash = rest. Calculate gross and net calorific value of fuel. Assume latent heat of steam to be equal to 587 cals/gm.

- Define fuel. Give the classification of fuels with suitable examples.
- A gaseous fuel has the following composition by volume, $CH_4 = 40\%$, $H_2 = 40\%$, $C_2H_6 = 6\%$, $N_2 = 1\%$. Calculate the volume of air required for complete combustion.
- What is cracking? Differentiate between catalytic cracking and thermal cracking.

May 2004

- Name the highest ranking coal and give its approximate calorific value.
- Explain the production of ethyl alcohol from molasses.
- 3 g of coal was heated in Kjeldahl's flask and NH_3 gas evolved was observed in 40 ml of 0.5 N H_2SO_4. After absorption, the excess acid required 18.5 ml of 0.5 N KOH for exact neutralisation. 2.3 g of coal sample in quantitative analysis gave 0.35 g $BaSO_4$. Calculate the percentage of N and S in coal sample.
- Explain cetane number.
- A gaseous fuel has the following composition by volume – $CH_4 = 35\%$, $CO = 10\%$, $H_2 = 6\%$, $C_3H_8 = 10\%$. Calculate the weight of air required for complete combustion.
- Explain proximate analysis of coal.

Dec. 2004

- Name a few characteristics of a good fuel.
- Write the method for estimation of percentage of any two of the following:
  (1) Percentage of volatile matters
  (2) Percentage of carbon and hydrogen
  (3) Percentage of sulphur
- A gaseous fuel has the following composition by volume.
  $H_2 = 35.00\%$, $CH_4 = 47.00\%$, $C_2H_6 = 4.00\%$, $CO = 6.00\%$, $N_2 = 8.00\%$.
  Calculate the volume of air required for complete combustion of one cubic metre of this fuel.
- Write short notes on H.C.V. and L.C.V.
- A sample of coal contains $C = 61\%$, $O = 32\%$, $H = 6\%$, $S = 0.5\%$, $N = 0.2\%$, $Ash = 0.3\%$. Calculate high and low calorific value using Dulong Petit's formula.

May 2005

- What is the role of ash in coal?
- Select the compound which has the highest octane number and highest cetane number out of n-heptane, n-hexadecane, n-octane, iso-octane.

- What is the power alcohol? How is it obtained from molasses?
- A gas has the following composition by volume. $CH_4 = 40\%$, $C_2H_6 = 20\%$, $C_3H_8 = 10\%$, $H_2 = 10\%$, $CO = 10\%$, $N_2 = 10\%$. Calculate weight of air required per m^3 of gas. (Mol. wt. of air = 28.94)
- Explain octane rating and role of antiknock agents in gasoline.

Dec. 2005

- Name a few characteristics of a good fuel.
- Give a brief account of various types of coal. Write their composition and calorific value.
- Calculate the gross and net calorific value of coal having following composition:
  C = 80%, H = 7%, O = 3%, S = 3.5%, N = 2.1% and ash = 4.4%.
- What is cracking? Distinguish between thermal and catalytic cracking.
- A gas has the following composition by volume $H_2 = 20\%$, $CH_4 = 6\%$, $CO = 22\%$, $CO_2 = 4\%$, $O_2 = 4\%$ and $N_2 = 44\%$. Find the volume of air actually supplied per m^3 of this gas.

May 2006

- What is cracking? Explain the advantages of catalytic cracking over thermal cracking.
- Explain how sulphur can be estimated in coal.
- A sample of coal has the following composition by mass C = 70%, O = 8%, H = 10%, N = 3%, S = 2%, Ash = 7%. Calculate H.C.V. and L.C.V. using Dulong's formula.
  (**Ans:** G.C.V. = 8805.8 Kcal/Kg, L.C.V. = 8277.5 Kcal/Kg)
- What is petroleum? How is crude petroleum refined? Name various products obtained from it?
- Calculate the weight of air needed for complete combustion of 1 gm of coal containing:
  C = 72%, H = 10%, O = 9%, N = 3% and remaining being ash.
- Explain knocking and role of antiknocking agents in gasoline.

Dec. 2006

- A sample of coal contains C = 61%, O = 32%, H = 6%, S = 0.5%, N = 0.2% and Ash = 0.3%. Calculate its G.C.V. and N.C.V. by Dulong and Petit's formula.
  (**Ans:** G.C.V. = 5630 Kcal/Kg, L.C.V. = 5313.02 Kcal/Kg)

- 0.2 gm of coal sample is accurately weighed and is burnt in a combustion apparatus. The gaseous products of combustion are absorbed in potash bulb and calcium chloride tubes of known weight. The increase in weight of potash bulb and $CaCl_2$ tubes are 0.66 gm and 0.08 gm respectively. Calculate the percentage of carbon and hydrogen in the coal sample.
- Discuss fixed bed catalytic cracking. What are the advantages of catalytic cracking over thermal cracking?
- Discuss in brief, the production of alcohol by fermentation.
- The composition of gas was found to be H_2 = 10%, CH_4 = 16%, C_2H_6 = 20%, N_2 = 6%, CO = 22%, CO_2 = 18%, O_2 = Rest.

  Calculate the volume of air required for complete combustion of 10 m^3 of this gas.

May 2007

- What is meant by octane number? Explain the role of antiknocking agents.
- 1.56 gm of the coal was Kjeldahlized and NH_3 gas thus evolved was absorbed in 50.0 ml of 0.1 N H_2SO_4. After absorption the excess (residual) acid required 6.25 ml of 0.1 N NaOH for exact neutralization. 2.60 gm of the coal sample in a qualitative analysis gave 0.1755 gm of $BaSO_4$. Calculate the percentage of N and S in the coal sample.
- Describe the refining of petroleum with various fractions obtained.
- Calculate the volume of air required for complete combustion of 1 m^3 of gaseous fuel which possess by volume:

  CH_4 = 35%, C_2H_4 = 5%, CO = 15%, H_2 = 40%, N_2 = 1%. Water vapour = 4%.
- A sample of coal contains C = 70%, O = 20%, H = 8%, S = 0.5%, N = 0.5%, Ash = 1.0%. Calculate the gross and net calorific value of Dulong and Petit's formula. (H.C.V. = 7564.7 kcal/kg, L.C.V. = 7142.06 kcal/kg)

Dec. 2007

- Define (i) Octane number (ii) Cetane number. Write the significance of each.
- How the percentage of volatile matter is determined during proximate analysis of coal? What is its significance?
- A sample of coal contains:

  C = 60%, O = 33%, H = 6%,

  S = 0.5%, N = 0.3%, Ash = 0.2%.

  Calculate the gross and net calorific values.

- A sample of coal has the following percentage composition by mass:
  C = 70%, H = 10%, O = 5%,
  N = 2%, S = 3% and remaining ash.
  Calculate the minimum amount of air needed for complete combustion of 1 kg of coal.
- How alcohol is manufactured from sugar cane by fermentation process?

May 2008

- Define fuels? Classify fuels with suitable examples. (3)
- Define term calorific value and ignition temperature. (3)
- The % composition of coal is as follows:
  C = 80%, H = 6%, O = 8%, S = 1.5%, N = 1% and Ash = rest.
  Calculate gross and net calorific value of fuel. (5)
- What is cracking? Write advantages of catalytic cracking over thermal cracking. (5)
- What is biodiesel? Explain the method to obtain biodiesel from vegetable oil? Give advantages of biodiesel. (5)
- A sample of coal contains C = 65%, H = 13%, O = 6%, S = 4%, N = 2% and Ash = remaining.
  Calculate minimum amount of air needed for complete combustion of 1 kg of coal. (5)
- Explain refining of petroleum with suitable diagram. (5)

CHAPTER 4

GREEN CHEMISTRY

INTRODUCTION

Environmentally benign chemical synthesis which reduces the use and generation of hazardous substances is commonly known as Green Chemistry or clean chemistry. Advanced countries like USA strictly follow the procedures for various syntheses so as to reduce or eliminate hazardous by-products which are responsible for the pollution of the environment. The chemists are motivated for the environmentally benign synthesis of new products and also to develop green synthesis for existing chemicals by the replacement of the organic solvents, which are hazardous or eliminate the use of solvent altogether.

By synthesis of drugs, etc. the green chemistry has brought about medical revolution. The food supply of the world has increased due to the discovery of hybrid varieties, improved methods of farming, better seeds and use of fertilizers, insecticides and herbicides, etc. The living standard, of life has improved due to the discovery of dyes, plastics, cosmetics and other materials. The most important development effects are release of hazardous by-products of chemical industries and agro chemicals in the atmosphere which are responsible for polluting the environment. Green chemistry is highly effective approach to prevent environmental pollution.

4.1 THE NEED OF GREEN CHEMISTRY

The developments in the 20th century brought about various benefits to the mankind, but most of them were responsible for a number of environmental problems at the local and global levels. Hence, it is necessary to formulate guidelines and pass strict rules, bring about changes at the grass root level.

Necessary changes in the chemistry curriculum in the schools, colleges and the universities are needed to achieve the desired effect at various levels about green chemistry.

4.2 PRINCIPLES OF GREEN CHEMISTRY

Green chemistry deals with any chemical process or technology which improves environment and quality of our life. The objectives are to minimise environmental pollution by developing harmless chemical process. A set of these principles are formulated by Anastas and Paul T as follows.

4.2.1 Prevention of Waste

Always it is better to prevent waste than to treat or clean up waste after it is formed. It is best to carry out a synthesis by designing a pathway so that formation of waste or by-products is minimised or avoided. The cost of the treatment and disposal of waste adds to the overall cost of production. It must be kept in mind that 'prevention is better than cure'. The disposal of the waste in the atmosphere, not only causes pollution but also requires expenditure for cleaning up.

4.2.2 Maximisation of Atom Economy

The synthesis design of the product should be such that all the materials/chemicals used in the process are converted into the product. On utilisation of materials considered waste and causes disposal problems which are not economical.

If the yield of a reaction is about 91%, it is considered to be good. The percentage yield is calculated by

$$\text{Percentage yield} = \left[\frac{\text{Actual yield of the product}}{\text{Theoretical yield of the product}}\right] \times 100$$

If one mole of a starting material produces one mole of the product, the yield is 100%. However, there may be generation of significant amount of waste or by-products in such synthesis which is not visible in the above calculation. Even though is 100% yield in such synthesis is not considered to be a green synthesis. For example, in Grignard and Witting reaction yield is 100% but large amount of by-products obtained they do not take into account.

A reaction or a synthesis is considered to be green if maximum of the strating materials or reagents are converted in the final product. The percentage atom utilization is determined by the following equation:

Percentage atom utilization =

$$\frac{\text{MW of desired product}}{\text{MW of desired product} + \text{MW of waste products}} \times 100$$

Trost B.M. developed a concept of atom economy that how much of the reactants are converted in the final product. Sheldon R.A. also developed this concept as follows:

$$\text{Percentage atom economy} = \frac{\text{FW of atoms utilized}}{\text{FW of the reactants used in the reaction}} \times 100$$

(FW = Formula wt.)

The most common reactions are rearrangement, addition, substitution and elimination reactions. Let us find out which of the following reactions is more atom economical.

Maleic anhydride production routes

(1) 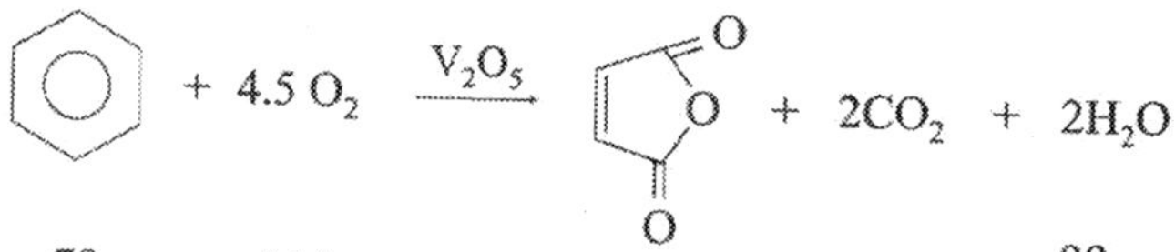

78 144 98 $= \frac{98}{78 + 144} = 44.1\%$

(2) 

$= \frac{98}{152} = 64.5\%$

Butene

(3) 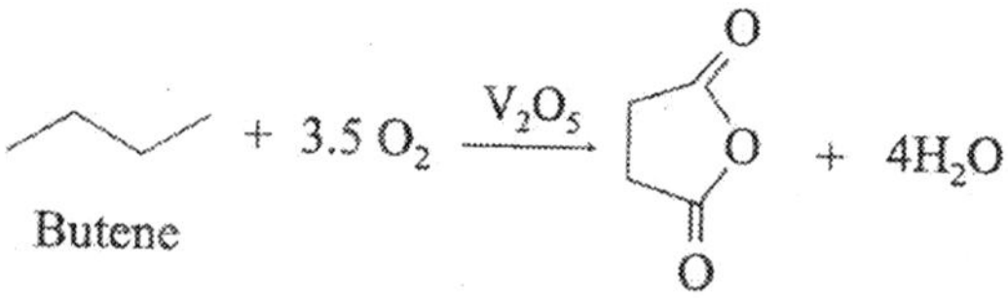

Butene

(a) Rearrangement Reactions

Allyl phenyl ether on heating at 200°C gives o-allyl phenol.

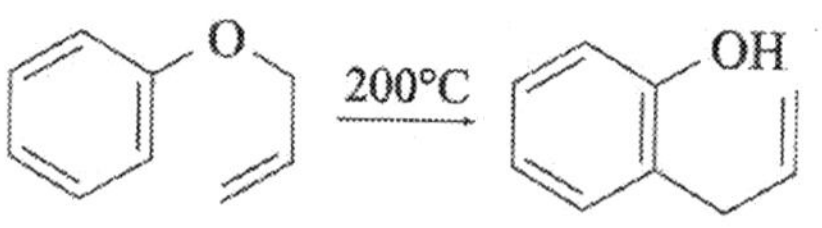

Allyl phenyl ether O-allyl phenol

100% atom economical reaction all the reactants are incorporated to product.

(b) Addition Reactions

For example, bromination of propene

(i) $H_3C\,CH = CH_2 + Br_2 \xrightarrow{CCl_4} H_3C\,CH\,Br\,CH_2Br$ (100% atom economical)

Propene 1,2-dibromopropane

(ii) 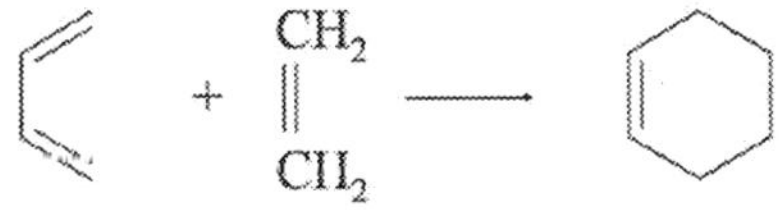 (100% atom economical)

Butadiene Ethene Cyclohexene

(iii) $H_3C{-}CH = CH_2 + H_2 \xrightarrow{Ni} H_3C - CH_2 - CH_3$ (100% atom economical)

Propene Propene

(c) Substitution Reactions

For example:

$$CH_3CH_2\overset{O}{\overset{\|}{C}}OC_2H_5 + H_3CNH_2 \longrightarrow CH_3CH_2\overset{O}{\overset{\|}{C}}NHCH_3 + CH_3CH_2OH$$

Ethyl propionate Methyl amine N-methyl propamide Ethyl alcohol

In the above reaction, OC_2H_5 is not incorporated in the amide and one hydrogen atom of the amine is not utilised.

	Reactants		*Utilised*		*Unutilised*	
	Formula	*FW*	*Formula*	*FW*	*Formula*	*FW*
	$C_5H_{10}O_2$	102.132	C_3H_5O	57.057	C_2H_5O	45.061
	CH_5N	31.057	CH_4N	30.049	H	1.008
Total	$C_6H_{15}NO_2$	133.189	C_4H_9NO	87.106	C_2H_5OH	46.069

Therefore, the percentage atom economy $= \dfrac{87.106}{133.189} \times 100 = 65.40\%$.

(d) Elimination Reactions (Hofmann elimination)

In these elimination reactions, two atoms or groups of atoms are lost fom the reactant to form a π bond.

(i) 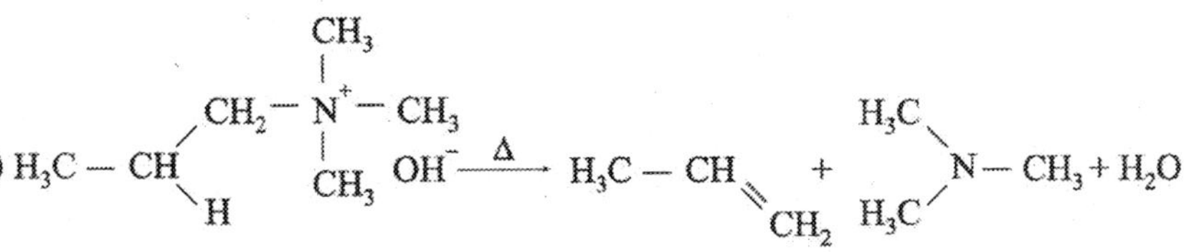

The above elimination reaction is least economical as percentage atom economy is only 35.30%.

(ii)

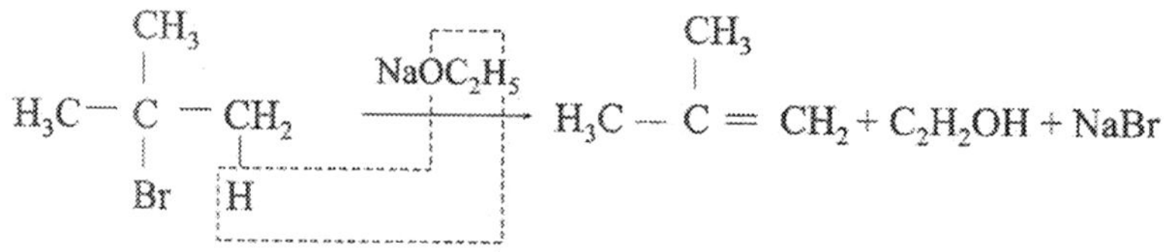

2-Bromo-2-methylpropane 2-Methyl propene

The percentage atom economy is less than the Hofmann elimination reaction, only 27%.

4.2.3 Chemical Synthesis (Non-hazardous)

Synthetic methodologies should be designed to generate substances without toxicity to human health and the environment.

Important principle of Green Chemistry is to prevent or at least minimise the formation of environmentally harmful products. In case these are formed, their effects on the workers must be minimised by various techniques. The expenditure will add to the cost and if control fails there may be more risk involved. For example, leakage of MIC an intermediate in the manufacture of agricultural pesticide at Bhopal in 1984 which was highly poisonous. Hence, generate substances with little or no toxicity to humans and environment is recommended by Green Chemistry.

(i) Synthesis of aromatic amines

Environmentally hazardous chlorinated aromatics are known to be persistant bioaccumulators. Stern and coworkers proved that use of chlorinated aromatics can be avoided by using nucleophilic substitution for hydrogen.

(ii) Synthesis of indigo

Indigo, the dye used to colour blue Jeans can be prepared enzymatically by removal of the side chain of tryptophan to give indole, which can be dehydroxylated enzymatically and the deoxudised with oxygen to indigo. The presently used commercial process starts with aniline which is highly toxic. Considerable amount of waste salts, thereby causes disposal problems.

For production of indigo is carried in more greener route in America. The yield is 5 times more than traditional plant.

A. Greener route (using enzymatic transformation)

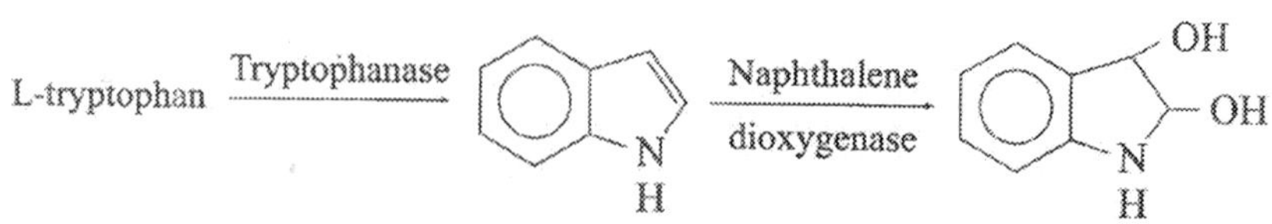

B. Conventional route (Using hazardous aniline)

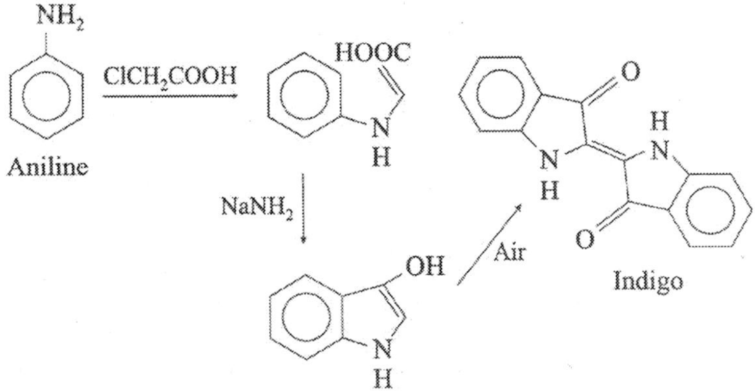

4.2.4 Safer Design of Chemicals and Products

Chemical products should be designed to preserve efficiency of function while reducing toxicity.

The synthesised products like dyes, paints, cosmetics, pharmaceuticals, etc. should be safe to use. Thalidomide introduced in 1961 for reducing the effects of nausea and vomitting during pregnancy was found to be unsafe. The children of women taking thalidomide suffered from birth defects, hence the use of thalidomide was banned. Strict regulations were passed for testing all new drugs.

Many insecticides like DDT, aldrin, gamaxane are also found to be toxic to human beings.

4.2.5 The use of Auxiliary Substances (Solvents, Separating Agents, etc.)

Due to the excellent solvent properties methylene chloride, chloroform, perchloroethylene, carbon tetrachloride, benzene and other aromatic hydrocarbons have been used in a large number of reactions. The halogenated solvents have been identified as carcinogenous, benzene and other aromatic hydrocarbons are cancer promoters. For dry cleaning of the fabrics perchloroethylene was used which highly toxic is now replaced by liquid CO_2 nowadays.

The solvent used in a reaction should not cause any environmental pollution and health hazard. As far as possible in aqueous phase or without the use of a solvent in solid phase reaction should be carried. Chlorofluorocarbons (CFCs) widely used as cleaning agents, blowing agent and as refrigerants, are responsible for depleting the ozone layer, which in turn has disastrous effect on human survival. Compounds like carbon tetrachloride, methylene chloride, chloroform, etc., which have been used as solvents have disastrous effects in the atmosphere. The clear air act in some of the advanced countries like USA to control many classes of chemical used as solvents have been made.

4.2.6 Energy Requirement should be Minimised by Considering their Environmental Impacts

Energy efficiency of a process can be increased by proper heat transfer and avoiding wastage of energy during the process. This can be achieved by the use of catalyst and proper designing of the process.

4.2.7 A Raw Material or Feedstock Should be Renewable

The starting materials may be obtained from renewable or non-renewable material. Petrochemicals are mostly obtained from petroleum oil, which is a non-renewable source. Agricultural or biological processes can be used to obtain renewable starting materials; however, continuous supply due to the factors like failure of crops, etc. will not be possible. Substances like carbon dioxide and methane gas are considered renewable starting materials as they are available in abundance.

Synthesis of adipic acid

New biosynthetic pathway have been reported for the manufacture of adipic acid by Frost at Michigan State University by using catechol and hydroquinone using glucose as starting material instead of traditionally used benzene.

Traditional Pathway

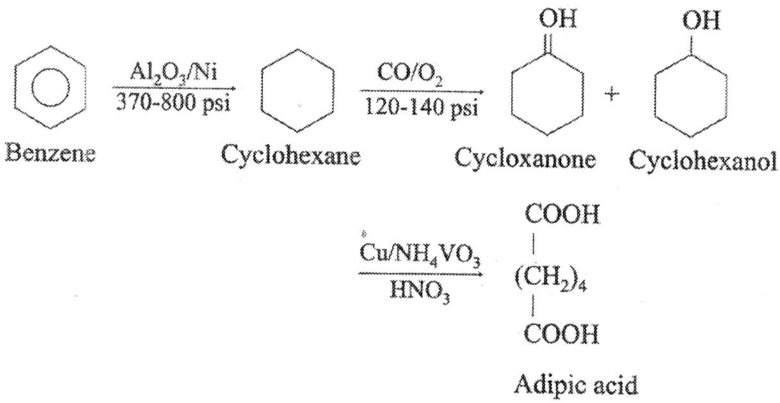

Greener Pathway (absolute safe)

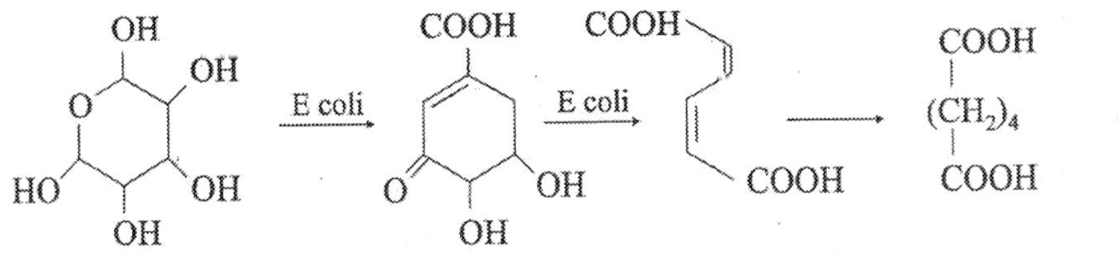

D. Glucose 3-dyhydro-sikimate Cis-Muconic acid Adipic acid

4.2.8 Avoid Unnecessary Derivatisation

As far as possible blocking, protection and temp. modification should be avoided.

Derivatisation increases steps of the process and additional reagents are required. Example, benzyl chloride (a known hazard) needs to be handled with care.

Traditional Synthesis of Ibuprofen (Atom economy-40%)

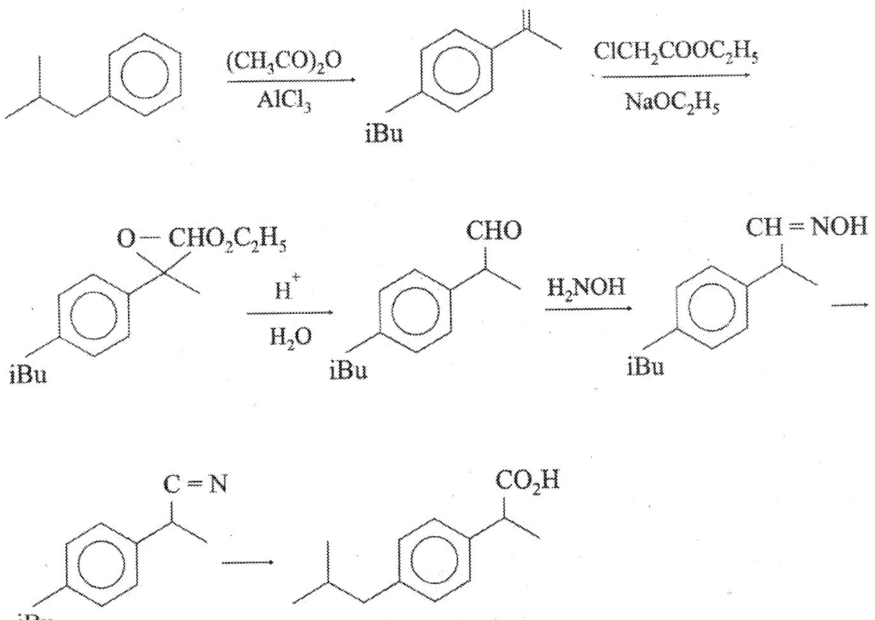

Ibuprofen

Alternative Synthesis of Ibuprofen (Atom economy 77%)

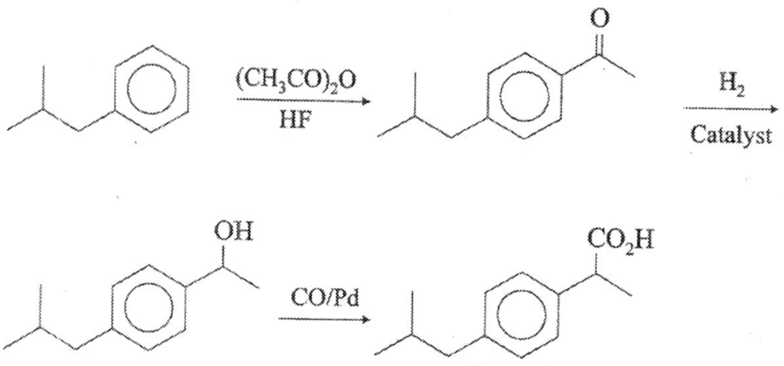

Ibuprofen

4.2.9 Use of Catalytic Reagents

- Hydrogenation of olefins: It is observed that catalytic reactions are faster hence requires less energy and minimises waste. Recent processes have been developed which use high toxic recoverable catalysts. Examples:

$$\underset{\text{Propene}}{H_3C{-}CH{=}CH_2} + H_2 \xrightarrow{Ni} \underset{\text{Propane}}{H_3C{-}CH_2{-}CH_3}$$

- Synthesis of benzyl cyanide from benzyl chloride by using phase transfer catalyst.

$$\underset{\text{Benzyl chloride}}{C_6H_5CH_2Cl} + \text{aq. KCN} \xrightarrow{PTC} C_6H_5CH_2CN$$

> 90% yield

Benzyl cyanide

(PTC-phase transfer catalysis)

- Synthesis of benzoic acid from toulene.
- Hydration of acetylene to give acetaldehyde or ketones.

$$\underset{\text{Acetylene}}{HC{\equiv}CH} + H_2O \xrightarrow[H_2SO_4]{HgSO_4} \underset{\text{acetaldehyde}}{CH_3CHO}$$

4.2.10 Proper Designing of the Chemical Products

Designing should be such that at the end of their function they do not persist in the environment.

The DDT was one of the first pesticides which gets accumulated in soil, plants and animals for many years causing pollution. Hence, the products synthesised should be biodegradable and should not be persistent materials. Plastics, polystyrene are also non-biodegradable causes pollution, an alternative plastics containing cellulose may be used.

4.2.11 Analytical Methodologies for Monitoring and Control the Process Prior to the Formation of Hazardous Substances

The prevention and minimisation of the generation of hazardous substances in chemical processes may be controlled by using sensors, monitors and analytical techniques, e.g. perfectly monitoring preparation of ethylene glycol can avoid production of hazardous materials.

4.2.12 Minimisation of the Potential for Accidents

The accidents in Bhopal and several many other cities have resulted in the loss of thousands of lives. Hence, occurrence of accidents in chemical industries must be avoided by using safer chemicals using optimum temperature, pressure and other working conditions. Prevention of solvents to escape in atmosphere must be avoided. Catalysts are also helpful in reducing potential of the accidents.

4.3 INDUSTRIAL APPLICATIONS OF GREEN CHEMISTRY

The green chemistry finds wide applications in many chemical processes like extraction and manufacture of fats, oils, pigments, pharmaceutical compounds, polymers, biotechnology, etc.

Oleochemicals (USA) offers complete range of chemical products. Bio-refineries in Germany is first developing bioactive polymer (Biopos). From the cornwaste 1,50,000 tonnes of ecoplastic is annually produced in US, which can be used for routine packaging like foil or plastic pots and even T-shirts.

4.4 GREEN SOLVENTS

Alternative non-hazardous organic solvents emerged by supercritical media. Supercritical fluids are also used in production of polymers. In the production of Teflon using supercritical CO_2 instead of trichlorotrifluoroethane has developed by using dopant. In nano electronics controllable evaporation sales can be obtained by using supercritical fluids like CO_2 which can replace ultra high purity water without affecting nanometric structures.

Ionic liquids (ionic solvents) are used as green solvents with the advantage of low vapour pressure. A commercial process by generation of ionic liquid has been developed in Germany by Ludwig Maase at BASF which also absorbs unwanted by-product HCl and hence it is the most effective.

4.5 GREEN FUELS AND E-GREEN PROPELLANTS

In 2006 Prof. Galen J. Suppes, University of Columbia was awarded an academic award for converting waste glycerine into biodiesel. For space industry also an alternative fuel known as "Green propellants" are needed. The H_2O_2 and H_2 are the better alternatives. H_2O_2 decomposes without producing any harmful compounds.

4.6 BIOCATALYSIS

In an alternative greener process the corrosive acid catalysts are replaced by zeolites (hydrated sodium aluminotsilicates), caesium zeolite catalyst were found to be very effective in base catalysis.

This includes enzyme, antibody and biometric catalysis, enzyme catalysis widely used in biotechnology. Enzyme catalysis is not only applicable to aqueous media but for non-aqueous and supercritical fluids also.

Degussa in 1981 developed a concept of enzyme membrane, "Projektas catalyse" center established by him. With the help of biocatalysis several greener pathways for medicine and other compounds are established.

Synthesis of acryl amide

(1) Conventional method

$$\underset{\text{Acrylonitrile}}{CH_2 = CH - CN} + H_2O \xrightarrow[2\,NH_3]{H_2SO_4} \underset{\text{Acryl amide}}{CH_2 = CH - \overset{\overset{O}{\|}}{C} - NH_2} + (NH_4)_2SO_4$$

(2) Green pathway

$$CH_2 = CH - CN \xrightarrow[\text{Nitril}]{\text{Hydritase}} \underset{\text{Acryl amide}}{CH_2 = CH - \overset{\overset{O}{\|}}{C} - NH_2}$$

4.7 SIGNIFICANCE OF GREEN CHEMISTRY

- Green chemistry helps industry to develop cleaner and non-hazardous chemical processes.
- Waste utilisation.
- Green organic synthesis avoids waste and thereby reduces disposal problems.
- It also reduces requirement of energy.

- Renewable materials are utilised.
- By using catalyst production cost of the materials is minimised.
- Use of improper solvents avoided which helps in keeping the environment clean.

QUESTIONS

- Define Green Chemistry. Explain objectives of green chemistry.
- Explain the principles of Green Chemistry with suitable examples.
- Why should we avoid derivatisation in reaction?
- Explain the benefits of maximising atom economy.
- How prevention of waste is dangerous?
- Discuss the applications of green chemistry.
- Calculate % atom economy of the following reactions:

(i) 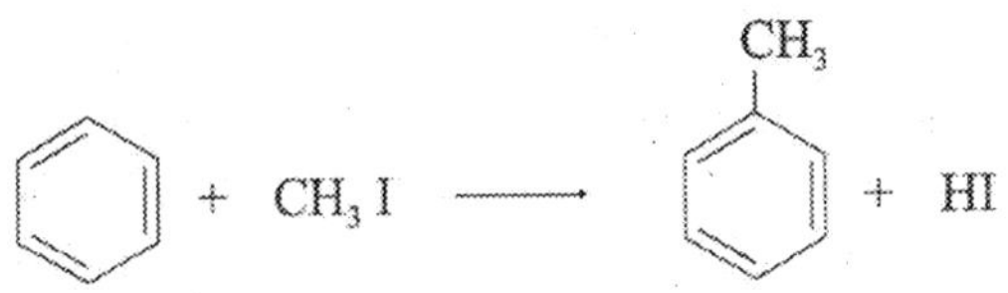

(ii) $C_2H_5Br + KOH \xrightarrow{H_2O} C_2H_5OH + KBr$

(iii) $C_3H_7Cl + KOH \xrightarrow{C_2H_5OH} CH_3{-}CH{=}CH_2 + KBr + H_2O$

MUMBAI UNIVERSITY QUESTIONS

May 2008

- What is Green Chemistry? List principles of Green Chemistry. (3)
- Explain basic idea in the field of green chemistry research with the help of synthesis of indigo dye. (5)
- Calculate % atom economy for the formation of maleic anhydride reaction. (3)

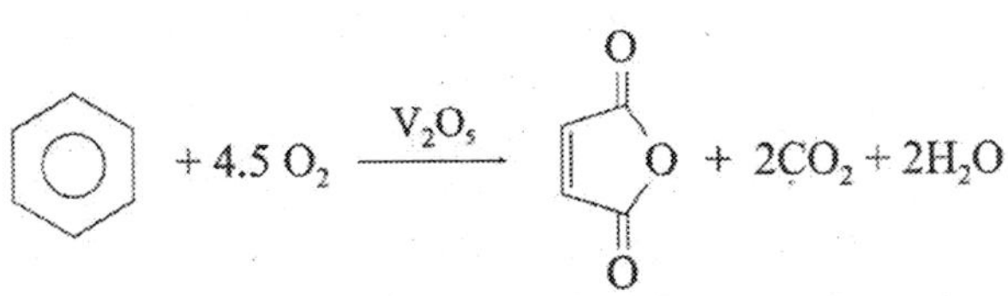

Benzene Maleic anhydride

- Write a note on Green solvents. (2)

CHAPTER 5

COMPOSITE MATERIALS

5.1 INTRODUCTION

In the early periods of history, mud and clay were used for building walls of the houses for several years. Later on, it was realised that mats of bamboo, grass or cane impart higher strength, safety and durability to the mud walls. The development of which has showed better property and composite idea was developed. Composites such as asbestos, reinforced concrete is accepted universally. High performance composite materials like carbon fibre, glass fibre reinforced plastic composites, rubber based composites, synthetic fibres, steel cords, metal fibres, ceramic items developed recently.

The composite wood is defined as a macroscopic combination of two or more distinct materials having an identifiable interface between them. Composite exhibits unique properties and better characteristics as compared to its constituents. It can be also defined as combination of two or more materials fibrous or particulate reinforced supported by binder or matrix materials. In a composite two phases not only exist but act together so as to produce the desired characteristics. The demand of the aerospace community promoted development in reinforced composites of light weight and high strength Al-alloys. In architect and construction engineering the demand of composite material is continuously increasing because no homogeneous structural material provides desired attributes for a particular demand. In composite the fibre is required to support all main load. The mechanical capability depends on effectiveness of the reinforcement—which can be enhanced by incorporation of reinforcing fibre like glass fibre or cross-linking. The discovery of light weight, strong, high performance composites is very effective in construction and aerospace. Synthetic fibres like Kevlar, aeromatic resins as matrix materials are suitable for designing of high performance FRP composites.

5.2 MATRIX MATERIALS

The resins used are epoxy, phenolic, vinyl ester silicones and polyamide. The maximum working temperatures are 175°C for dry epoxies and 125°C for wet one. Vinylester have been used in Kevlar type of fibres. Recently thermosetting polyamide used within working temperature 350°C. The thermoplastic resin PEEK (polyether ether ketone) gives better toughness and moisture resistance.

These bind the fibres together and act as the medium by which external stress is transmitted and distributed to the fibres. It also protects the individual fibres from surface damage by mechanical abrasion or chemical reactions with the environment. Matrix separates the individual fibres and prevents the propagation of brittle cracks from fibre to fibre, which may result in catastrophic failure. Thus, this phase serves as a barrier to crack propagation. The matrix material should be ductile with lower elastic modulus. If the adhesive bonding forces between the fibre and matrix are high, the fibre pull-out is minimised. Bonding strength is most important in matrix-fibre combination. Ultimate strength of a composite mainly depends on the magnitude of this bond. As ductility is essential, only metals like Al, Cu and commercial polymers are used as the matrix materials.

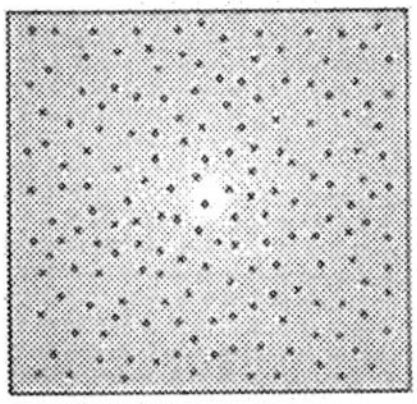

(a) Particle reinforcement in polymer matrix

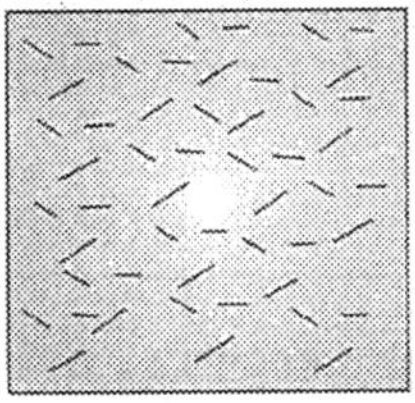

(b) Fibre reinforcement in polymer matrix

Fig. 5.1

5.3 FIBRE MATERIALS

Dispersed phase comprises fibres particularly of small diameter as they have higher strength than the bulk materials. These are very thin single crystals possessing extremely large length/diameter ratios. These are exceptionally strong with very high degree of crystalline perfection. Whiskers are the strongest well known materials, their use is prohibited as they are costly and the process of incorporating them in matrix phase is tedious. Aluminium oxide, silicon nitrides and carbides, graphite are the examples of whiskers. The polycrystalline and amorphous polymers with small diameter are preferred.

Some important dispersed phase constituents.

- **Glass fibres:** Green fibres are popular reinforcing materials which have very high tensile strength, thermal stability and good toughness. These are produced by passing glass melt through small orifice which is then cooled. 20 to 40% glass fibres are used to fill the polymers with coupling agents like methacrylic chromic chloride complex or organosilane.

- **Graphite or carbon fibres:** Oxidation of carbon under pressure and at low temperature followed by carbonisation at 1000°C these fibres are produced. These are strong and stiff even at high temperature. They have good resistance to corrosion. The thermal conductivity of graphite fibres is fairly high. These are used in aerospace and sporting goods.
- **Aramid fibres:** It is a aromatic polyamide with liquid crystal property. These have very high tensile strength, good resistance and thermal stability, e.g. Nomex and Kevlar, originally these were produced for tyre cords now they are used for filament bound rocket motor cases and pressure vessels. These are used in interior cabin structures in aircraft and military helmets.
- **Particles:** Metal oxides, carbide and nitrides, carbon black, mica, etc. can be used to obtain the composite. These have higher mechanical resistance and better thermal and electrical conductivity.
- High carbon steel, molybdenum, tungsten wires are also used as disphersal phase. Due to their larger diameter in wire form, they are more suitable for automobile tyres.

5.4 COMPOSITE WOOD MATERIALS

These are also known as wood based fibres. Binding material and additives are added to wood to produce required properties. These are classified as fibre board, particle boards and plywoods.

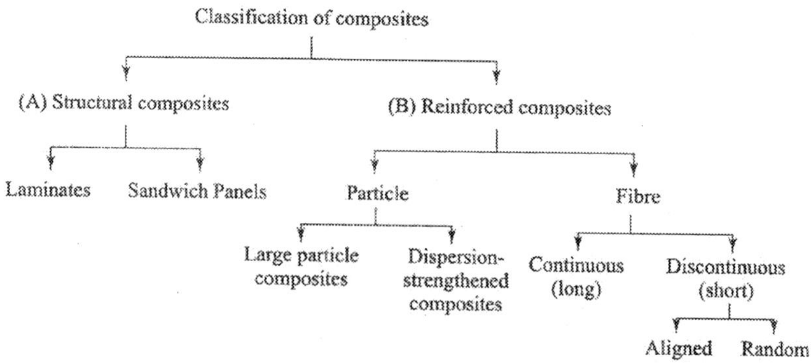

5.5 STRUCTURAL COMPOSITES

The properties of these composites depend not only on the properties of constituent materials but also on the geometry of the composites. These are broadly classified as (i) Laminar composites, and (ii) Sandwich panel composites.

5.5.1 Laminar Composites

It consists of two-dimensional panels or sheets which possess preferred high strength directions. Plywood is an aluminated composite containing thin layers of wood and continuous aligned fibre reinforced plastics are in preferred directions. Laminar composites can also be constructed using fabric materials such as paper, glass fibre, cotton in a suitable plastic matrix. They possess high strength in a number of directions in the two-dimensional planes. These composites are generally strong in both directions of reinforcement. They are used in furniture making, ceiling for diffuse lighting, interior in premises, etc.

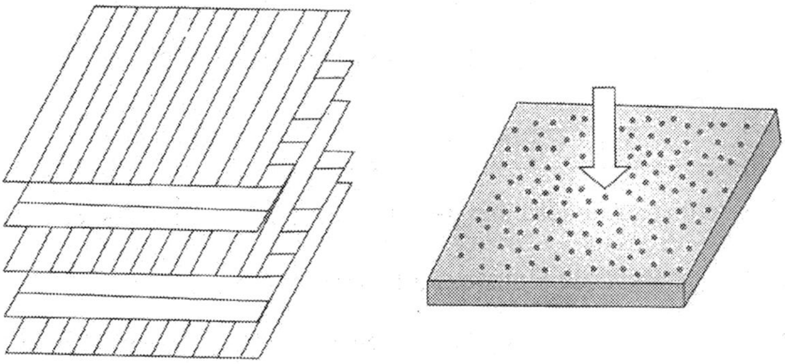

(a) (b)

Fig. 5.2 Oriented fibre reinforced layers successively stacked for producing a fabricated laminar composite.

5.5.2 Sandwich Panel

Another type of a composite consists of two strong outer sheets called faces which are separated by a layer of less dense material of lower strength called core. Fibre reinforced plastics, aluminium alloys, steels, titanium and plastics are used as face materials. The less dense materials (cores) includes inorganic cementing materials, foamed polymers, synthetic rubbers, etc. The faces and cores are joined together with the help of an adhesive. The faces bear most of the in-plane loading and transverse bonding stresses. The cores separates faces from each other and provides resistance along planes perpendicular to the faces.

"Honeycomb" is a popular type of core material structure of this contains thin foils forming interlocked hexagonal cells with their axes oriented at right angles perpendicular direction to the face sheet. Sandwich panels have low density, high tensile strength and elasticity, excellent dimensional stability and good corrosion resistance. They are extensively used in roofs, floor and walls of buildings, in aircraft for wings, fuselage and tail pane skins.

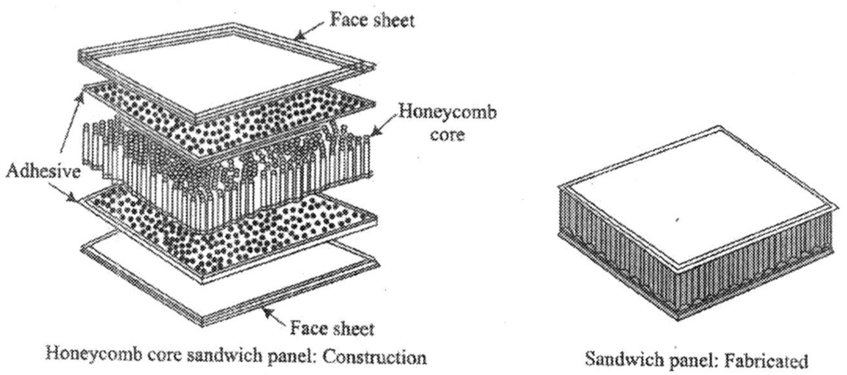

Fig. 5.3 Schematic diagram.

5.6 REINFORCED COMPOSITES

These are further subdivided into particle and fibre reinforced composites.

5.6.1 Large Particle Reinforced Composites

In these composites reinforced particles help to retain the movement of the matrix phase in the vicinity of each particle. The particulate phase is harder and stiffer than the matrix, which acts as a reinforcing material. The mechanical properties of the composites depends upon the bond strength between two phases.

Examples of particle reinforced composites are concrete and cermets. Concrete consists of matrix phase of cement with sand and gravels as particulate matter. RCC is harder than the ordinary cement. The ceramic metal composites are known as cermets. The most widely used is cemented carbide which contains tungsten or titanium carbide embedded in a matrix of cobalt or nickel. These find use as cutting tools for hardened steels. No single material can provide excellent properties like cermets, carbon black consisting of very small spherical carbon provides when added to vulcanised rubber enhance its tensile strength and toughness, 15-30% of carbon black is used for automobile tyres.

5.6.2 Dispersion Strengthened Composites

Smaller size particles 10 to 100 mm are generally used in dispersion strengthened composites. The mechanism of strengthening is similar to that of precipitation hardening commonly used for hardening high strength aluminium alloys which is also called 'Age hardening' because strength development is with respect to time. Alloys like Mg-Al, Al-Cu, Cu-Sn, Cu-Be and some ferrous alloys are hardened and converted into composite materials with ceramics. Fine metallic oxides and non-metallic particles are used to harden and strengthen metal and metal alloys. By addition of 3% (by volume) ThO_2 (thoria) in the form of finely dispersed particles, the high temperature strength of nickel alloys can be considerably enhanced. The materials are called thoria-dispersed nickel (TDN). In Al-Al_2O_3 system similar effect is produced.

5.7 FIBRE REINFORCED COMPOSITES

These materials contain dispersed phase in the form of fibres. These composites are important in providing high strength and stiffness on the weight basis. The characteristic property is known as high specific strength and high specific modulus.

$$\text{Specific strength} = \frac{\text{Tensile strength}}{\text{Specific gravity}}, \text{ Specific modulus} = \frac{\text{Modulus of elastic}}{\text{Specific gravity}}$$

When low-density fibres and matrix materials are used, high specific strength and high specific moduli composites are produced. Mechanical properties of fibre reinforced composites depend on properties of fibre and matrix, fibre length, fibre orientation and concentration.

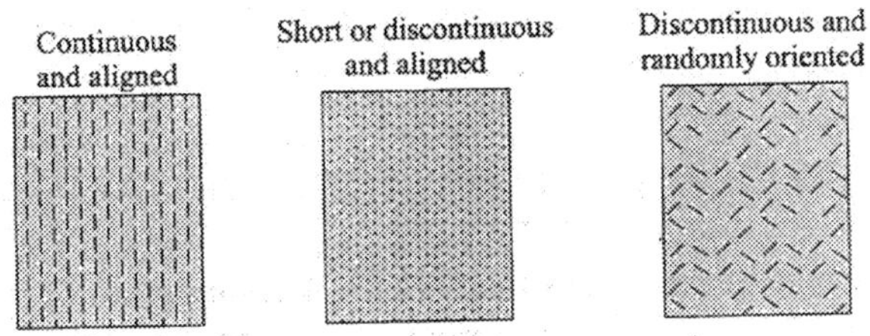

Fig. 5.4 Different types of fibres.

5.7.1 Continuous and Aligned

When fibre length is much greater then the fibres are termed continuous fibres. Their properties depend on the direction in which they are measured, i.e. anisotropic. Maximum reinforcement and strength is achieved along the longitudinal direction (aligned).

5.7.2 Discontinuous (short) Fibre Composites

These are further classified into discontinuous and aligned fibre composites and discontinuous and randomly oriented fibre composites. The discontinuous fibres efficiency of reinforcement is lower than the continuous. Even though they have greater commercial applications. Examples of fibre reinforced composites are:

- **Glass fibre reinforced polymer composites:** Consist of fibre phase glass fibres and matrix phase nylon, polysters, etc. They have low density and high tensile strength and higher resistance to corrosion and chemicals.

They are used in transportation industries, for storage tanks, automobile parts and pipes. As they have low stiffness they are used only at temperature.

- **Carbon fibre reinforced composites:** The fibre phase is carbon fibres and matrix phase of this composite is polymer like nylon, polysters, etc. These have high density, good dimensional stability even at high temperature and excellent corrosion resistance. They are used in aircraft, sports goods, etc. As their cost is high they are used only for specific purposes.
- **Aramid fibre reinforced composites:** These are short and long fibre composites. Shorter discontinuous ones consist of fibre phase and polymer matrix. They have high surface area, high resistance to wear and good heat stability. They are used in automobiles.
- Long or continuous ones are ductile and can sustain stresses. They are used in motors in aircraft. Their use is because of the high cost.
- Alumina and/or carbon fibre-reinforced metal composite contains aluminium oxide, carbon fibre or tungston as a fibre phase and Al, Ni and Co-based alloys as a matrix phase. They have low density and excellent strength. Good resistance to wear, creep and thermal distortion. They are used in automobile industries and to make components of turbine engine.

5.8 APPLICATIONS OF COMPOSITE MATERIALS

Composites are widely used in electronics and telecommunication, transportation, mobiles, automobiles, sports good, agriculture, construction, etc.

QUESTIONS

- What are composites? Give suitable examples with their classification.
- Explain the terms (i) matrix (ii) dispersed phase.
- Explain the following:
  - (i) Ar amide fibre reinforced polymer composites
  - (ii) Dispersion-strengthened composites
  - (iii) Large particle reinforced composites
  - (iv) Laminar composites
- What are the applications of composite materials?
- Explain the sandwich panel composites with suitable examples.
- What are the main functions of matrix phase and particulate phase of composites.
- Explain the cermets with suitable examples.

- Write a note on fibre glass reinforced composites.
- What are whiskers? Explain their properties with examples.
- What are the characteristics of matrix and particulate phase.

MUMBAI UNIVERSITY QUESTIONS

May 2008

- Write the applications of composite materials. (3)
- Define a composite material? Give the classification of composite material? Explain laminates. (5)
- Write notes on the following: (4)
  - (i) Matric phase of composite material
  - (ii) Cermets
- What are the characteristic properties of composite materials? (3)

CHAPTER 6

CATALYSIS

6.1 INTRODUCTION

As mentioned earlier one of the simple method to increase the rate of reaction is to increase the temperature of reaction. This may lead to many undesirable effects. On the other hand, catalysts speed up reaction without increasing the temperature. A catalyst is a substance which alters the rate or the velocity of a chemical reaction, without undergoing any change in it. The phenomenon of altering the rate of a chemical reaction by use of a catalyst, is known as catalysis. A catalyst which accelerates the rate of reaction is called positive catalyst and the catalyst which retards the rate of reaction, is called negative catalyst.

Positive Catalysis

- In certain reactions the rate of reaction increases by addition of a catalyst. For example, decomposition of $KClO_3$ in the presence of MnO_2 (rate increases). The catalysts are not able to initiate the reaction or ultimate products of the reaction do not alter in catalytic reaction. Only in the presence of MnO_2 decomposition becomes very fast.

$$2\,KClO_3 \xrightarrow[\text{heat}]{MnO_2} 2\,KCl + 3O_2$$

- Biological reactions are enhanced by enzymes and hydrolysis of an ester is enhanced by mineral acids as positive catalyst.

Negative Catalysis

- In certain reactions, the rate of reaction decreases by addition of a catalyst. For example, decomposition rate of H_2O_2 into H_2O and O_2, is decreased by the presence of dilute acid or glycerol.

- Rate of oxidation of $CHCl_3$ is decreased in the presence of alcohol as a catalyst.

$$2CHCl_3 + O_2 \xrightarrow{\text{Alcohol}} 2COCl_2 + 2HCl$$

- Rate of atmospheric oxidation of alkaline sodium bisulphite is reduced considerably by addition of small amount of alcohol.

$$Na_2SO_3 \xrightarrow{\text{atm}} Na_2SO_4$$

$$Na_2SO_3 \xrightarrow[C_2H_5OH]{\text{atm}} Na_2SO_4\text{(slow)}$$

- Promoters and Poisons: Efficiency of a catalyst may be altered by the presence of even small quantity or traces of certain other substances. A promotor is the substance which increases the activity of a catalyst. In the manufacture of NH_3 from N_2 and H_2, iron acts as catalyst but the Mo or $(Al_2O_3 + K_2O)$ in small quantity along with iron, acts as a promotor.
- Anticatalyst or catalytic-poison is the substance which reduces or destroys the activity of the catalyst when present in a small quantity. Presence of even traces As_2O_3 in the spongy platinum catalyst inhibits the catalytic action of platinum in the manufacture of H_2SO_4.

6.2 CHARACTERISTICS OF CATALYSIS

- A catalyst remains unchanged chemically at the end of the reaction.
- A small amount of catalyst is sufficient.
- The equilibrium constant of a reversible reaction is not altered by a catalyst.
- Specific action of a catalyst.
- Activity of a catalyst is altered by promotors and an anticatalyst.
- A catalyst remains unchanged chemically at the end of reaction.

  The chemical composition of a catalyst remains the same. However, a physical change like colour, size of particles, etc. may change at the end of the reaction, e.g. platinum gauze used as catalyst, may become rough, after some time during the oxidation of NH_3.
- A small amount of a catalyst is sufficient to alter a chemical reaction, in many cases. The results of a finely powdered catalyst are better, because of larger surface of area of catalyst in fine powdered form. Catalysts are not consumed in a chemical reaction. Catalyst used while initiating reaction is generated at the end of the reaction. However, in many homogeneous catalytic reactions, rate is also dependent on the amount of the catalyst used and increases with increase in concentration of a catalyst.

In many homogeneous reactions rate of reaction increases with increase in surface area of the catalyst. Hence, finely powdered catalysts are better in these reactions.

- The equilibrium constant of a reversible reaction is not altered by a catalyst.

In a reversible reaction a catalyst catalyses both the forward and backward reactions, and hence does not change the equilibrium state.

However, the catalyst helps to attain the equilibrium state quickly without altering equilibrium constant of the reversible reaction,

$$N_{2(g)} + 3H_{2(g)} \underset{}{\overset{Fe}{\rightleftharpoons}} 2NH_{3(g)}$$

At the given temperature the equilibrium constant of reaction between N_2 and H_2, remains the same even if Fe catalyst is used.

- *Specific action of a catalyst:* Catalysts are specific in nature thus same reactant can give different products in the presence of different catalysts. For example, reaction of CO and H_2 can lead to different products in the presence of different catalysts.

$$CO + 3H_2 \xrightarrow{Ni} CH_4 + H_2O$$

$$CO + H_2 \xrightarrow{Cu} HCHO$$

$$CO + 2H_2 \xrightarrow{ZnO + Cr_2O_3} CH_3OH$$

- The cracking of a higher alkane is catalysed by ($Al_2O_3 + SiO_2$) catalyst. The alkane may be n-heptane or iso-ocatane or dodecane.
- There is a particular catalyst for a certain reaction. All the biological catalysts (enzymes) are highly specific in action.
- Activity of a catalyst is altered by promoters and anticatalysts.

In many cases activity of a catalyst can be significantly increased by the addition of small amounts of certain substances called promotors. These increase activity of catalyst but do not act as a catalyst. e.g. manufacture of NH_3 by Habers process

$$N_2 + 3H_2 \xrightarrow[(Mo)]{Fe} 2NH_3 \text{ (Mo-promotor)}$$

In hydrogenation of unsaturated oils nickel is used as catalyst to make saturated oil. Activity of nickel is increased by using small amounts of copper or tellurium.

The activity of a catalyst is completely destroyed by the presence of certain substances and these are called poisons.

In manufacture of sulphuric acid by combination of SO_2 and O_2 to give SO_3, the presence of arsenic oxide in small traces reduces the activity of the platinum catalyst.

6.3 THE CATALYSED REACTIONS

These are broadly classified into:

(i) Homogeneous catalysis

(ii) Heterogeneous catalysis

(iii) Enzyme catalysis

Three general types of catalysis reactions are:

(i) *Homogeneous catalysis*: When the reactants, products and catalysts are all in the same phase as the reaction mixture. Hydrolysis of ester is a common example in which small amount of acid catalyses the reaction to form acid and alcohol. Ethyl acetate on hydrolysis in the absence of acid.

$$CH_3COOC_2H_5 + H_2O \xrightarrow{k} CH_3COOH + C_2H_5OH$$

Rate of reaction $= k[CH_3COOC_2H_5]$ – First order kinetics

However, in the presence of small amount of HCl as a catalyst

rate of reaction $= kc\ [CH_3COOC_2H_5]\ [H^+]$

The H^+ ion attacks the $>C = 0$ group of ester using an oxygen lone pair, species formed is more susceptible to attack of water which helps in quickly breaking of ester to give acid and alcohol.

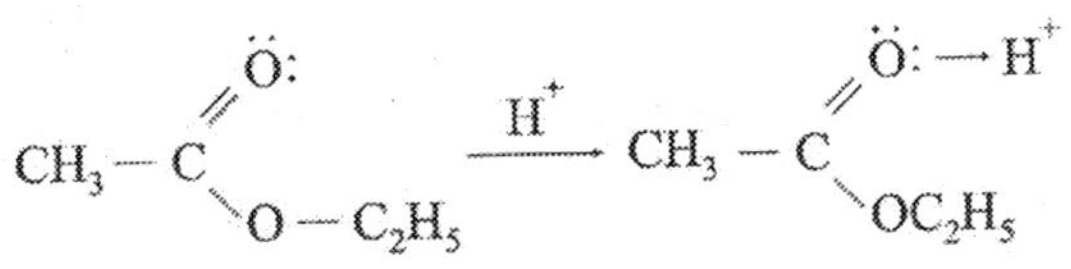

Another example of base catalysed homogeneous catalysis in gas phase is the conversion of SO_2 to OSO_3 in lead chamber process of sulphuric acid manufacturing as follows

$$S + O_2 \rightarrow SO_2$$
$$2SO_2 + O_2 \rightarrow 2SO_3$$
$$SO_3 + H_2O \rightarrow H_2SO_4$$

The rate of oxidation of SO_2 to SO_3 is low in the absence of a catalyst. If NO_2 is used as a catalyst, the reaction is more effective and proceeds at a faster rate.

$$2SO_2 + 2NO_2 \rightarrow 2SO_3 + 2NO$$
$$2NO + O_2 \rightarrow 2NO_2$$
$$\overline{2SO_2 + O_2 \rightarrow 2SO_3.}$$

There is no net loss of NO_2, reactant, product and catalyst are in the gas phase.

(ii) *Heterogeneous catalysis*: In this type of catalysis the reactant and products are either liquid or gases whereas the catalyst is solid. The phase of a catalyst is different from the reactant and products. A common example is formation of NH_3 by reaction of N_2 and H_2 gases. At room temperature reaction is very slow. At higher temperature the reaction becomes faster but decomposition of ammonia to nitrogen and hydrogen takes place. Fritz Haber in 1905 found that use of Fe and oxides of potassium and aluminium at 450°C is useful to produce ammonia on a large scale.

Another well-known example of catalytic converters used in automobiles to reduce harmful gases produced as exhaust.

$$2CO + O_2 \xrightarrow{\text{catalyst}} 2CO_2$$

$$2C_8H_{18} + 25O_2 \xrightarrow{\text{catalyst}} 16CO_2 + 18H_2O$$

$$2NO \xrightarrow{\text{catalyst}} N2 + O_2$$

These conversions are only possible by using catalysts which are heterogeneous as the reactants and products are gases.

(iii) *Enzyme Catalysis:* An enzyme is defined as a complex organic catalyst by living cells. A high molecular weight protein of complex structure with unique conformations at relatively small places. These unique conformations of enzyme are known as active sites of enzymes. Most of the enzymes are made up of amino acids.

General formula is $H_2N - \underset{\displaystyle R}{\underset{|}{CH}} - COOH$ (R = different—for different—amino acids)

for e.g. —CH_3(Alanine), —CH_2COOH (Aspartic acid), —H(glycine),—CH_2 —$CH(CH_3)_2$ (Leucine), $- \underset{\displaystyle CH_3}{\underset{|}{CH}} - C_2H_5$ (Isoleucine), —CH_2—CH_2—$CONH_2$ (glutamine),

Enzymes are capable of catalysing reactions at ~37°C (body temperature) and at physiological pH.

In human system most of the reactions are enzyme catalysed reactions.

Most of the enzymes are used in the cell which produces it and are called intracellular enzymes. When enzymes are used outside the cell which produces it are called extracellular enzymes.

When temperature is increased, the rate of enzyme catalysed reaction increases upto a certain temperature and then decreases. The temperature at which rate of reaction is maximum is called optimum temperature. At body temperature most of the enzymes show maximum activity at 0°C and 100°C the rate of enzyme catalysed reactions is practically zero and hence sterilisation of medical objects is carried out by boiling them in water

Enzyme catalysed reactions are sensitive to pH, highly acidc or alkaline pH cause denaturation of proteins. Enzyme reactions rate is maximum in the pH-range 6.0 to 8.00 except gastric juices.

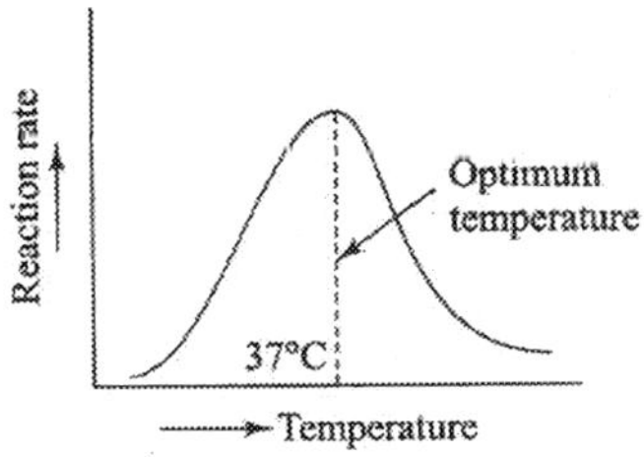

Fig. 6.1

When all other parameters of an enzyme catalysed reaction are kept constant and concentration of enzyme is varied, the rate of the reaction is increased as shown in the graph.

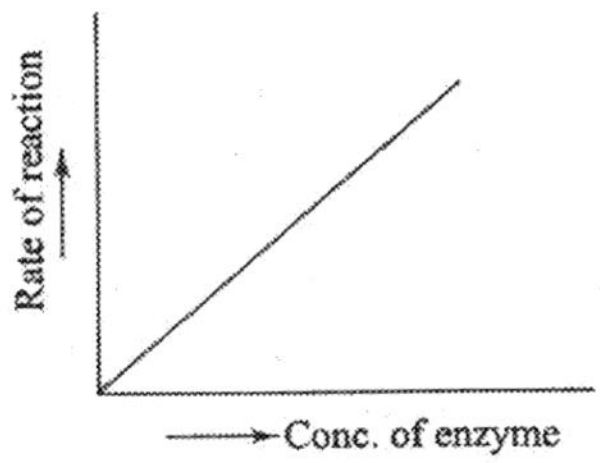

Fig. 6.2

The specificity of enzymes can be classified as

(a) **Absolute specific:** Enzymes that catalyse a particular reaction and have no effect on closely related substrates are called absolute specific. e.g

$$NH_2CONH_2 \xrightarrow[H_2O]{\text{Urease}} 2NH_3 + CO_2$$

Only hydrolysis of urea to NH_3 and CO_2 is catalysed by enzymes. It does not hydrolyse methyl or phenyl urea.

(b) **Group specific:** These act on structurally similar molecules, e.g. enzyme pepsin hydrolyses all peptide bonds (—CO—NH_2) having an adjacent aromatic amino acids.

(c) **Stereo specific:** When only one stereoisomer is catalysed by enzymes are called stereo specific enzymes. These have very high degree of specificity. For example, enzyme fumarase adds water molecules only to fumaric acid not to cis isomer maleic acid.

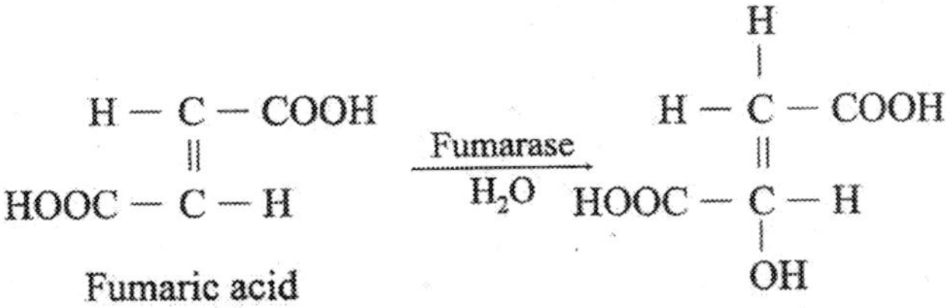

Fumaric acid

(d) **Linkage specific:** These are least selective in nature as they act on a particular type of linkage. Other structural features or groups near the linkage do not stop enzymes to act on the linkage. For example, lipase catalyses hydrolysis of ester linkage in lipids.

6.4 ACTIVATION ENERGY AND CATALYSTS

As per the collision theory for rates of chemical reactions, the rate of reaction depends upon the number of collisions of reacting molecules and the fraction of the colliding molecules with certain minimum additional energy (activation energy) so that the collision results in reaction.

The activation energy is the additional amount of energy to be supplied to the reactant molecules to reach the activated state (E_a). The reacting molecules absorb certain minimum energy to reach the activated state and form the unstable activated complex, which immediately decomposes to products, even though the reaction is exothermic or endothermic. More energy is required to supply for forming the activated state, if the activation energy of a reaction is larger for the exothermic reaction of H_2 and N_2 to form ammonia a high temperature is required, hence, there is need of a catalyst in such reactions to decrease the activation energy. Reactions which have higher E_a, require a suitable catalyst to reduce the activation energy, working temperature and pressure.

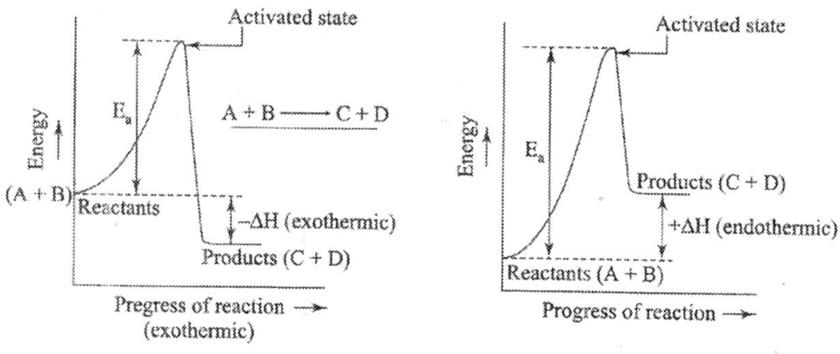

Fig. 6.3 Activation energies of reactions.

Reaction in which activation energy is smaller the reacting molecules may absorb energy from the surroundings, to reach the activated state of the reaction. The reaction between NH_3 gas and HCl gas occurs at room temperature or at lower temperatures, do not need to supply energy. The catalyst brings the reactants to come closer in the proper orientation to make effective collisions. They decrease the energy of activation barrier and hence reaction becomes faster. A typical reaction progress curve in the absence and presence of catalyst is shown in figure.

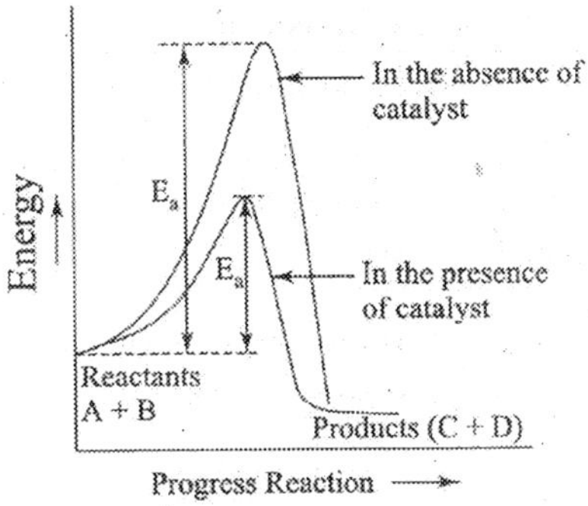

Fig. 6.4 Energy profile diagram for the reaction in the presence and absence of a catalyst.

- Arrhenius equation regarding relation between activation energy and rate constant of the reaction, is

$$K = A.e^{-Ea/RT}$$

K = rate constant of reaction

T = temperature of reaction in Kelvin.

R = gas constant (8314 kJ/mole)

E_a = activation energy

A = Arrhenius constant

- The activity of a catalyst is influenced by temperature, pressure and concentration

$$\therefore \quad Rate = Kf(P_i)$$

where K = apparent rate coefficient

P_i = partial pressure of reactant i.

The rate coefficient changes with temperature, pressure or surface concentration.

As per Arrhenius equation,

$$K = A' \exp(-E'/RT)$$

A′ = Temperature dependent pre-exponential factor.

E′ = Apparent activation energy of the reaction under study.

Only if the structure of a catalyst and concentration of the reactant remains unchanged E′ can be true activation energy but this condition is practically difficult, hence E′ cannot be true activation energy. Hence, defining of activity of catalyst in terms of activity energy concept is avoided and a new concept "Turn over frequency" (TOF) as Turnover number is used.

- TOF is defined as the number of times (n) that the catalytic reaction takes place per catalytic site per unit time for a fixed set of reaction condition, reactant ratio, extent of reaction. TOF measure of catalytic activity is successfully applicable in catalysts such as zeolites and finely powdered metal catalysts.

$$\text{TOF} = \frac{\text{Number of molecules of a given product}}{\text{Number of active sites} \times \text{Time}}$$

- "Reaction probability" (R_p) is another measure for catalytic activity. Catalytic hydrogenation, dehydrogenation, cracking reactions catalytic activity is calculated as follows.

$$R_p = \frac{\text{Rate of formation of products}}{\text{Rate of incidence of reactants}}$$

- In many chemical reactions, change of catalyst changes rate of reaction and also products. For example, oxidation of ethane.

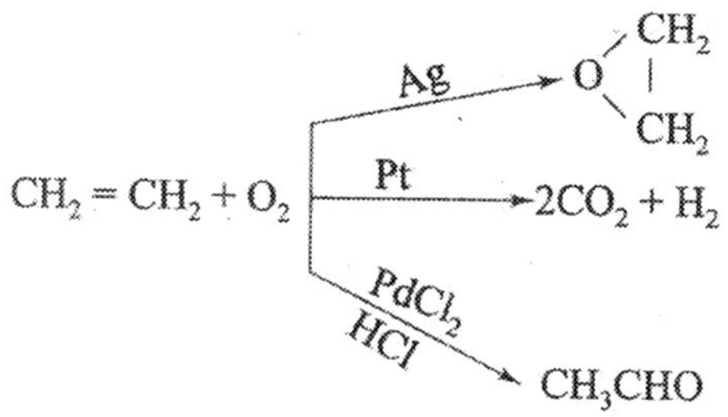

Most of the catalysts in heterogeneous catalysis, act through adsorption processes.

The reactions, which have high energy of activation are:

1. Manufacture of NH_3 and SO_3.
2. Cracking of larger hydrocarbons
3. Hydrogenation and polymerisation of vegetable oils and alkenes
4. Organic molecules synthesis by oxidation, reduction, hydro-desulphurisation, epoxidation, etc.
5. Biological reactions in living beings.

- High activation energy reactions either do not take place in the absence of catalysts or higher temperatures/pressures, are required to carry out reactions. By modification in manufacturing process, cost of material can be reduced. A catalyst decreases the activation energy of the reaction and makes the reaction possible at much lower temperature or pressure. e.g. $3H_2 + N_2 \rightarrow 2NH_3$ (Habers process), activation energy in absence of one catalysis 326 kJ/mol which decreases to ~150 kJ per mole in the presence of a catalyst. In reversible reaction, if the catalyst is used, the equilibrium condition can be reached much early. Many catalysts are generative hence only initial investment is required provided it is not poisoned.

Some important industrial processes using catalysts are given below:

- Wacker process

$$CH_2 = CH_2 + \frac{1}{2} O_2 \xrightarrow{PdCl_2 \text{ or } CuCl_2} CH_3CHO$$

- Ziegler-Natta plymerisation

$$nCH_2 = CH_2 \xrightarrow{TiCl_3 + Al(C_2H_5)_2Cl} -(-CH_2 - CH_2)_n$$

- Methanol from CO, H_2

$$2H_2 + CO \xrightarrow{ZnO + Cr_2O_3} CH_3OH$$

- Haber's process

$$N_2 + 3H_2 \xrightarrow{Fe + MO(promotor)} 2NH_3$$

- Contact process for H_2SO_4

$$2SO_2 + O_2 \xrightarrow{Pt \text{ or } V_2O_5} 2SO_3$$

- Ostwald process for HNO_3 (pt gause catalyst)
- Deacon's process for Cl_2($CuCl_2$ catalyst)
- Bosch process for H_2 ($Fe_2O_3 + Cr_2O_3$ catalyst)
- Cracking of hydrocarbons (Zeolites or $SiO_2 + Al_2O_3$ catalyst)
- Ethyl alcohol from sugar (Invertase, zymage catalyst)

6.5 THEORIES OF CATALYSIS

To explain catalytic action of a catalyst two theories have been proposed as under:

- **Theory of intermediate compound formation**

In 1806 Clement and Desormers proposed the formation of unstable intermediate compound during the catalysed reaction. According to the theory of intermediate formation, a catalyst combines with a reactant to form an unstable intermediate compound. This intermediate decomposes to give products and the catalyst is obtained back. Some typical reactions are as follows.

- Preparation of O_2 from $KClO_3$

$$2KClO_3 \xrightarrow{MnO_2} 2KCl + O_2$$

Actual reaction involves several steps.

$$2KClO_3 + 2MnO_2 \rightarrow 2KMnO_4 + Cl_2 + O_2$$
$$2KMnO_4 \rightarrow K_2MnO_4 + MnO_2 + O_2$$
$$K_2MnO_4 + Cl_2 \rightarrow 2KCl + MnO_2 + O_2.$$

The formation of $KMnO_4$ and Cl_2 is confirmed by several experiments. O_2 obtained by this method has been always found contaminated with chlorine gas and if the reaction flask is cooled suddenly at any stage a formation $KMnO_4$ is indicated by purple colour.

- Decomposition of formic acid to give CO and H_2O

$$HCOOH + H^+ \rightarrow (HCOOH_2)^+ \text{ (intermediate)}$$
$$(HCOOH_2)^+ \rightarrow H^+ + CO + H_2O$$

- An intermediate formation in enzyme catalysis have been characterised by spectral techniques. Common example is hydrolysis of sucrose by enzyme sucrase.

Sucrose + Sucrase ⟶ (Sucrose – Sucrase)
Intermediate

Sucrase + Glucose + fructose ⟵

- When ethanol reacts with H_2SO_4 there is a formation of ethyl hydrogen sulphate intermediate as follows.

$$C_2H_5OH + H_2SO_4 \rightarrow \underset{\text{ethyl hydrogen sulphate.}}{C_2H_5HSO_4} + H_2O$$
$$C_2H_5{\cdot}HSO_4 + C_2H_5OH \rightarrow C_2H_5OC_2H_5 + H_2SO_4$$

- Friedel-Craft alkylation reaction.

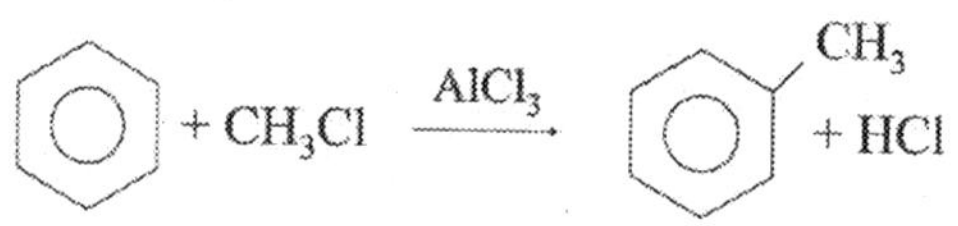

Mechanism $CH_3Cl + AlCl_3 \longrightarrow \underset{\text{Intermediate}}{[\overset{+}{C}H_3]\,[AlCl_4^-]}$

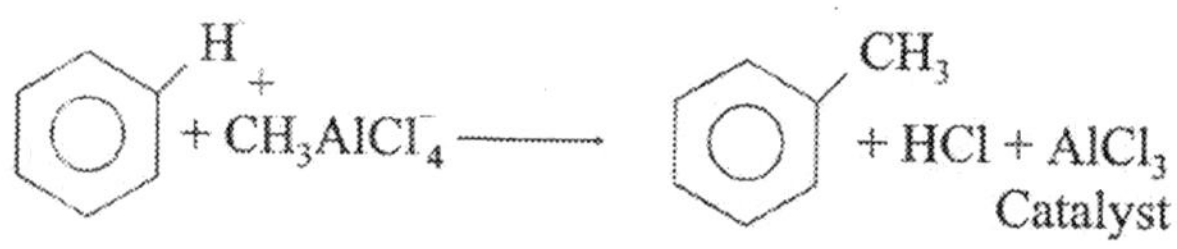

This theory does not explain the action of promotors and poisons on catalyst.

6.6 THEORY OF ADSORPTION

This theory is slightly similar to the enzymatic action according to which a catalyst surface has the same active points similar to the active sites of enzymes which hold significant quantities of liquid or gases. This phenomenon is called adsorption.

Adsorption may be physical or chemisorption.

- *Physical adsorption*: This occurs at the surface of catalyst due to van der Waal's forces. This adsorption is weak and the catalyst surface has no attraction for the gases. The adsorption of gas depends on the nature of gas and temperature. At low temperature adsorption is more which can be freed by altering the conditions. Physical adsorption increases with the increase in the surface area of the catalyst, hence finally divided catalysts are more efficient.

• **Chemisorption:** This occurs when there is some possibility of chemical action between reactant and a catalyst. For example, if the reactant is oxygen it can be adsorbed on metal by metal oxide formation and hydrogen can be hydride formation. Chemisorption occurs through valency forces with evolution of large amount of heat. Chemisorption is usually reversible at higher temperatures with increase in pressure it becomes more efficient.

For example, nickel catalyst is widely used in hydrogenation reactions. Hydrogen gas used is adsorbed at the surface of nickel metal. The covalent bond between two hydrogen atoms becomes weak and hydrogen atom binds to nickel atoms metallic surface. The chemical bond formed between nickel catalyst and hydrogen atom, the process is called chemisorption.

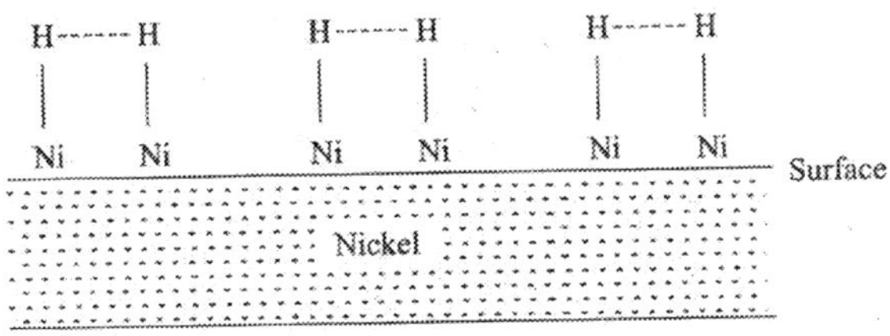

Fig. 6.5

Reaction of SO_2 with O_2 to give sulphur trioxide in which platinum catalyst is used adsorption of SO_2 and O_2 first occurs at the surface of platinum and then the molecules having sufficient energy makes bond with the surface causing chemisorption

$$SO_2 + O_2 \xrightarrow{pt} SO_3$$

An activated complex formed decomposes to give SO_3.

6.7 MOLECULAR DESIGN OF ZEOLITES

Hydrated sodium aluminosilicates, (Zeolites) are crystalline. They may be natural or synthetic.

- Synthetic zeolites, similar in structure to natural aluminosilicates. Fundamental building blocks of all zeolites are the tetrahedral of four oxygens having small Si atom or Al atom at the centre.

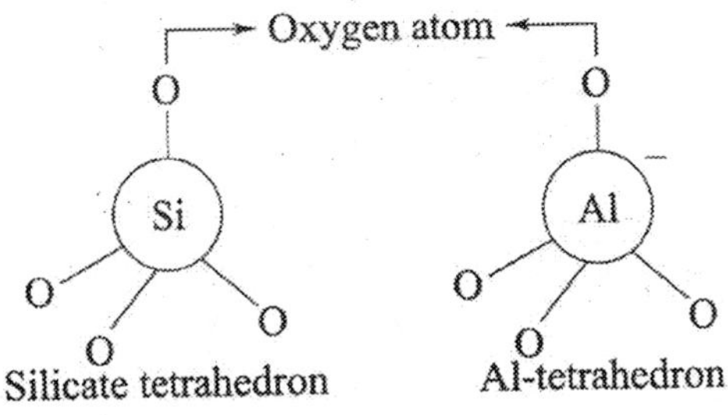

- There is negative charge on Al atom, in its tetrahedron because of 3 volency of Al. Si atom has 4 valency.
- In zeolites (aluminosilicates) oxygen atoms bridge between two silicon atoms or one slilicon and one aluminium atom.

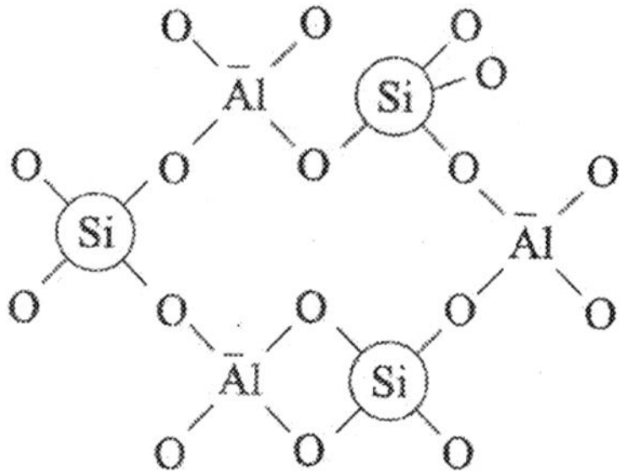

(Al & Si atom bridge with oxygen)

General formula of zeolite can be written as $Na_2O{\cdot}Al_2O_3{\cdot}xSiO_2{\cdot}yH_2O$

$$\text{Sodalite} = Na_6Al_6Si_6O_{24}{\cdot}2H_2O$$

$$\text{ZSM} - 5 = Na_3Al_3Si_{93}O_{192}{\cdot}16\ H_2O$$

$$\text{Mordenite} = Na_8Al_8\ Si_{40}O_{96}{\cdot}24H_2O$$

Sodalite cage structure is shown in Fig.6.4. Lines in structure represent oxygen bridges and corners represent silicon or aluminium atoms.

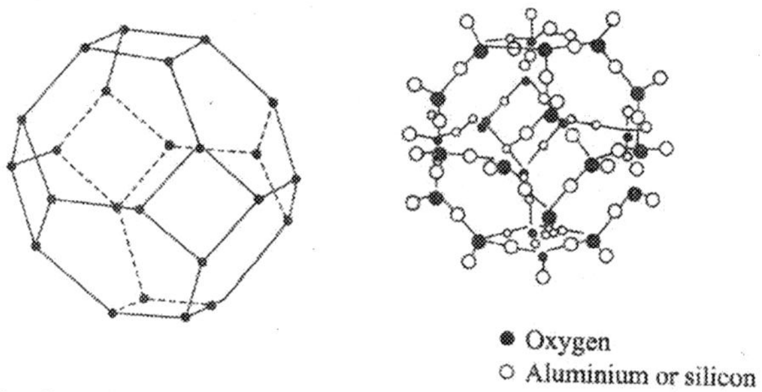

Fig. 6.6 Sodalite cage structure (truncated octahedron) O-atoms on the lines and Si or Al atoms at corners ($Na_6\,Al_6Si_6O_{24}\,.\,2H_2O$).

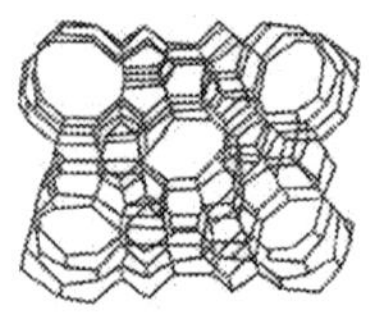

Fig. 6.7 ZSM-5 ($Na_3Al_3Si_{93}O_{192}16H_2O$) (Rings of 4, 5, 6, 7, 8, 10 size channel size 0.58 nm).

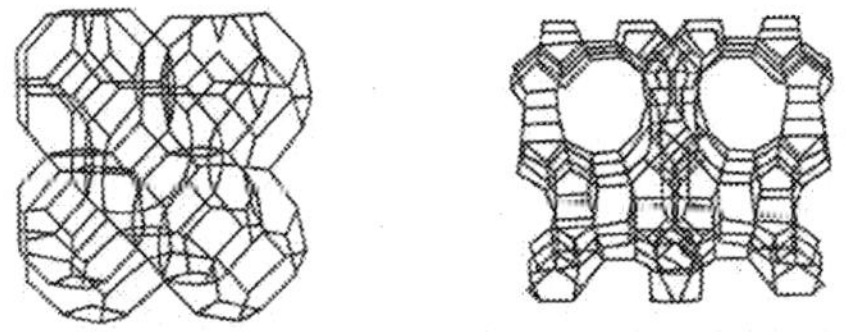

Sodalite
(Ring sizes 4 and 6
channel size 0.26 nm)

Mordenite ($Na_8Al_8Si_{40}O_{96}24H_2O$)
(Rings of 4, 5, 6, 8, 12 sizes
channel size 0.78 nm)

Fig. 6.8

Prepration zeolites: Zeolites are prepared by hydrothermal synthesis by using sodium aluminate, sodium hydroxide, sodium silicate and silicic acid.

Types of zeolite Structures: The nature of Zeolites prepared by hypothermal synthesis, depends on concentration of reactants, pH and temperature.

They differ in Si/Al ratio cation composition, cation exchangeability, thermal stability, adsorptivity catalytic character and in crystal lattice parameter.

Zeolite type	*Composition*				*Si/Al ratio*	*Free pore diameter*
	Na_2O	*SiO_2*	*Al_2O_3*	*H_2O*		
A	2	2	2	35	1.00	2.60A°
X	3.6	3	3 to 6	144	1 to 2	4.20 A°
Y	8	20	7 to 14	320	1.5 to 3	5.80 A°
Mordenite	6.3	27	5 to 7	61	4 to 6	6.70 A°

As the number of sodalite units in these building blocks are different they differ in size and shapes.

Zeolite surface structure: For sodium form, can be represented as,

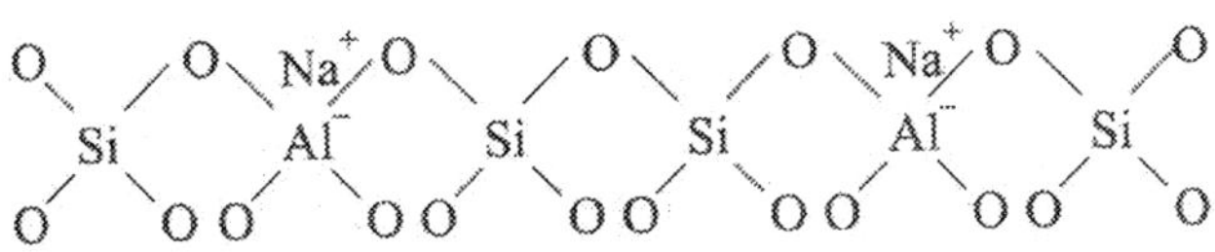

Important Properties of Zeolites

(i) Adsorption

- Because of free valancies on the surface of zeolite crystal, H^+, NH_4^+ can get adsorbed on the surface of zeolites. Smaller molecules of NH_3,CO, H_2O, etc., are also selectively adsorbed in the cavities of zeolites through H-bonding or due to electrostatic attraction with Al atom or sometimes due to formation of covalent bonds with the O-atoms on zeolite.

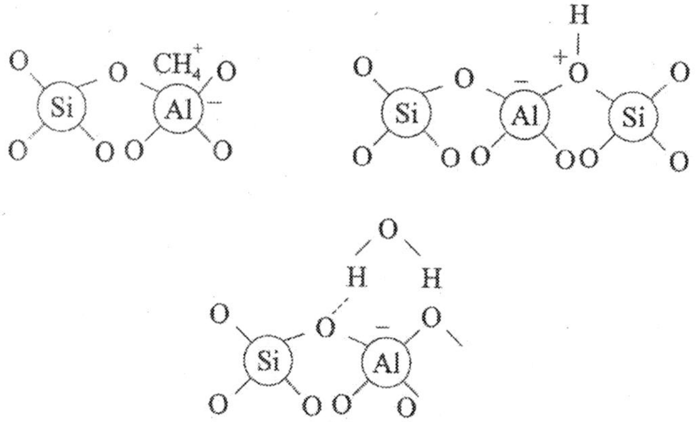

(ii) ***Catalysis***

- As zeolites are good adsorbants because of their fine microscopic structure. They act as catalysts for cracking, for isomerisation of hydrocarbons.
- Due to higher concentrations of active sites of zeolites they are preferred over silica or alumina. Zeolite can act as Lewis and Bronstead acid catalyst, in the reactions for formation of carbocations by polarisation in C–H bonds.
- Heterogeneous catalysis by acidic zeolite.

$$H_3C-OH+NH_3 \xrightarrow{\text{Acidic Zeolite}} H_3C-NH_2+H_2O$$

- Channel size favours methyl amino formation through the adsorption, rather than formation of di and trimethyl amines.

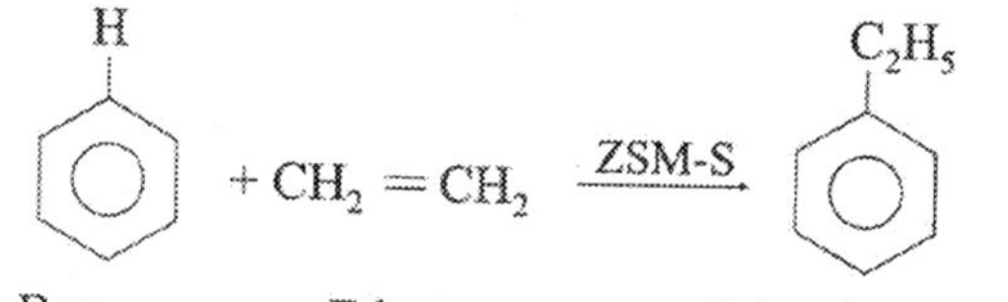

Benzene Ethyne Ethyl benzene

- ZSM-5 high performance catalyst used for manufacture of ethyl benzene has high acidity, peculiar shape, proper dimensions of channels.
- For redox catalysis a zeolite provides convenient framework sites for activation of transition metal ions Cu^{+2} exchanged zeolite, therefore, makes the conversion of NO_x pollutants in autoexhaust for conversion into N_2 and O_2 at high temperature.

$$2NO_x \xrightarrow[Cu^{+2}]{\text{Zeolite}} N_2+xO_2$$

(iii) Cation exchange

- Cations of smaller size can freely migrate through the open structure of zeolites, this property of cations exchange is utilised for softening of hard water.

6.8 METAL-COMPLEX CATALYSTS

Large number of organic manufacturing processes catalysed by metal complexes are homogenous catalytic reactions. For example, hydroformylation, dimerisation, oxidation, Ziegler-Natta polymerisation of alkenes, acetaldehyde from ethylene, viny acetate from ethylene and acetic acid, carbonylation of methanol, oxo process, etc. are carried out under mild conditions of temperature (upto 150°C) and pressure (20 atmosphere) when catalysed by metal complexes. The catalyst and reactants are dissolved in suitable organic solvent and the catalyst is recovered and recycled, after separation of the product by proper designing.

Chemical Structure of a Catalyst

Metal ions of Fe, Co, Pt, Pd, irridium, rhodium, etc. with the ligands having a completely filled p-orbital can from coordinate bond with the empty d-orbitals of transition metal ion. Some of the ligands are,

- **Neutral ligands:** $\ddot{N}H_3$, $H_2\ddot{O}$, $\ddot{C} \overset{\longleftarrow}{\equiv} O$ (with the lone pair of electrons) and $CH_2 = CH_2$ (and other alkenes, contain the π-bond electron cloud which can be donated to metal ion to form coordinate bond).
- **Negative ions ligands:** Cl^-, Br^-, I^-, CN^-, OH^-, etc. with two or more filled orbitals, which can overlap with empty d-orbitals of metal ion.

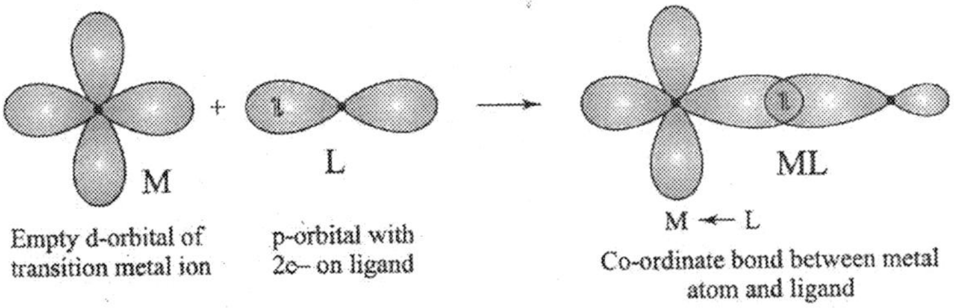

Fig. 6.9

The metal atom or ion is coordinately bonded with generally 4, 5, 6 or 8 ligand atoms known as coordination number.

The molecular structure of a metal complex with coordination number 4 (sp^3 or dsp^2 hybridisation) be either tetrahedral or square planar. Structure of metal complex with CN-6 (dsp^3 hybridisation) is trigonal bipyramidal or square pyramidal.

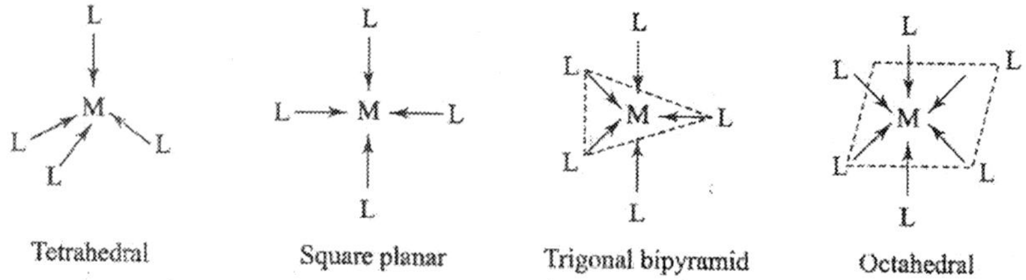

Fig. 6.10 Structure of metal-ligand complexes.

The structure of complex with C.N.6 (d^2sp^3 hybridisation) is usually octahedral or square bipyramidal.

The empty orbitals of metal atom in ns orbital, np orbital and (n – 1) d-orbital undergo hybridisation of different types and the shape of the complex molecule depends on its hybridisation.

The catalysis mechanism involves:

- replacement of weak ligand by strong ligand,
- conversion of ethylene to acetaldehyde (Wacker process)
- rearrangement of the organic ligand when H_2O ligand attacks to form unstable intermediate, which decomposes into product.
- the catalyst metal ion in the complex may get reduced to metallic state but can be easily recovered to complex state by reactions with ligand.

The mechanisms of some important metal – complexes catalytic processes.

$$H_2C = CH_2 + \frac{1}{2}O_2 \xrightarrow[110°C,\ 5\ atm]{[PdCl_4]^{-2}} H_3C - CHO$$

One ligand on the metal is replaced by another stronger lignad, during catalysis, by tetrachloropalladium ion $[PdCl_4]^{-2}$.

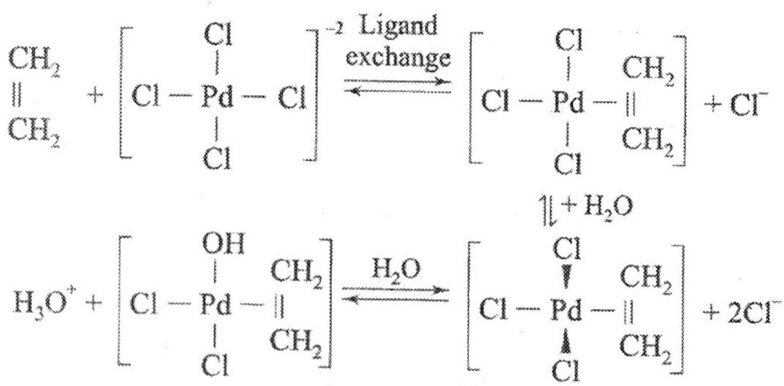

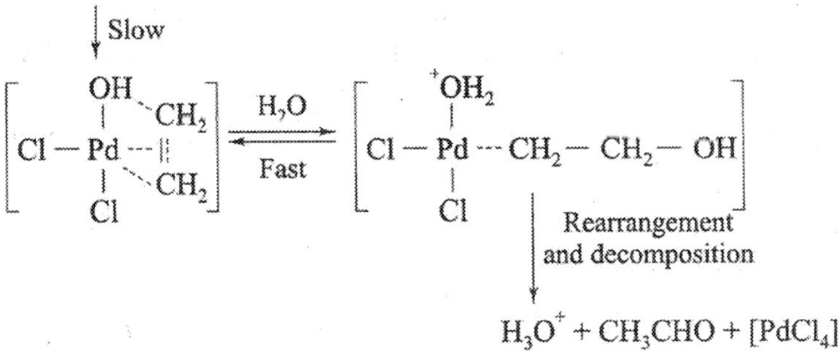

- **Methanol carbonylation** Ligand exchange mechanism, four co-cordinate to 6-coordinate complex formation, CO insertion and decomposition of complex to CH_3CHO. Promotor to the $Rh(CO)_4$ catalyst is CH_3I.

$$(CH_3OH + HI \rightleftharpoons CH_3I + H_2O)$$

$$Rh(CO)_4 + 2I^- \longrightarrow [Rh(CO)_2 I_2]^{-2} + 2CO$$

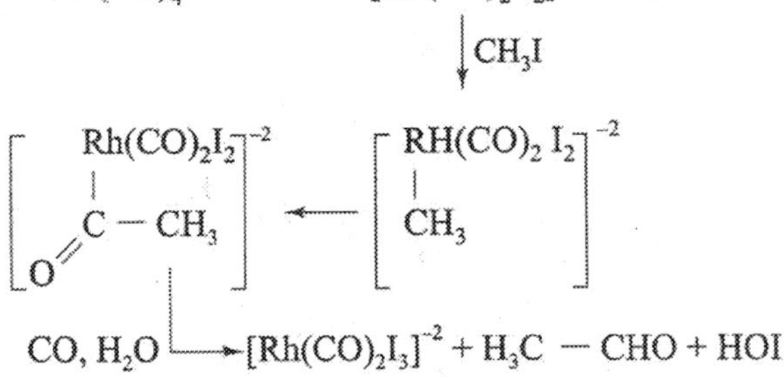

6.9 METAL CLUSTER CATALYSTS

The compounds containing metal-metal bonds are called as metal clusters, these are classified into

- Metal clusters containing polynuclear carbonyls, nitrosyls and related compounds of lower oxidation state of metal.
- Metal clusters containing chloride, oxide complexes of higher oxidation state of metal +2 to +3).

The formation of bond between the metal atoms is due to d-d overlapping of orbitals. The effective overlap is necessary to stabilise the metal cluster. When the metal atoms are at higher oxidation state (+2 to +3), the d-orbitals are contracted and overlapping is not effective.

Binuclear Compounds

- These are represented as $[M_2X_8]^{-2}$ ions, where M is the transition metal atom and X = Cl, Br, I or NCS, M = Mo, Re, Fe, Cu, Cr etc.

- For examples, in $(Re_2Cl_8]^{-2}$, the Re –Re distance is 2.24 A°. There is a quadruple bond between the metal atoms, one σ-bond, two π-bonds and one δ-bond.
- Each rhenium atom is bounded to four Cl atoms in a square planar array.

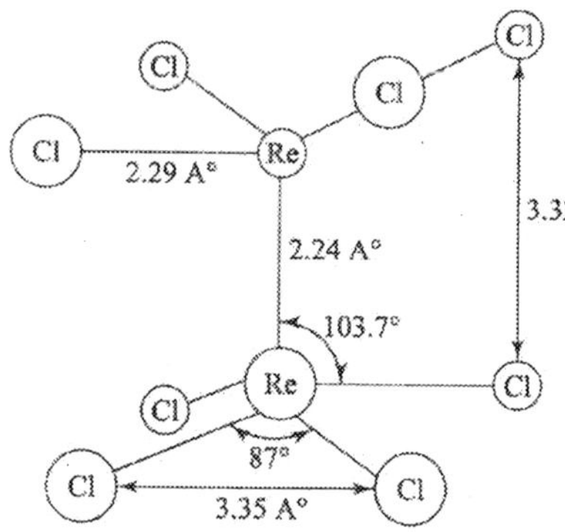

Fig. 6.11 Structure of $[Re_2Cl_8]^{-2}$

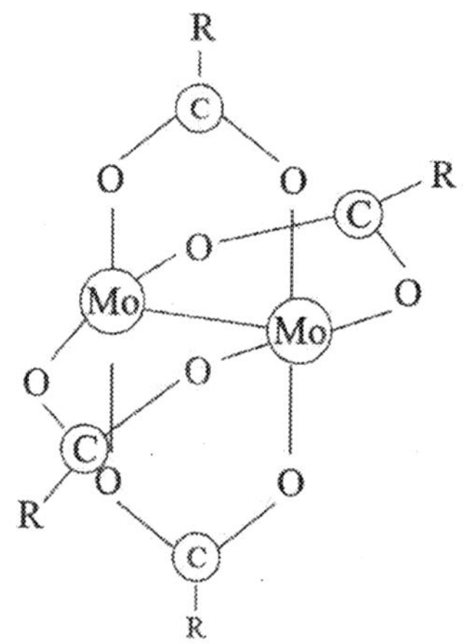

Fig. 6.12 Dinuclear carboxylate complex $Mo_2(RCOO)_4$

$[Mo_2Cl_8]^{-2}$ is the dinuclear metal cluster compound. Compounds which form a series of carboxylate complexes, $Re_2(RCOO)_2$ X_4, $Re_2(RCOO)_4$ X_2, $Mo(RCOO)_4$, etc. are also synthesised.

Trinuclear Clusters

Common examples are rhenium trihalides $[ReCl_3)_3$ and their derivations.

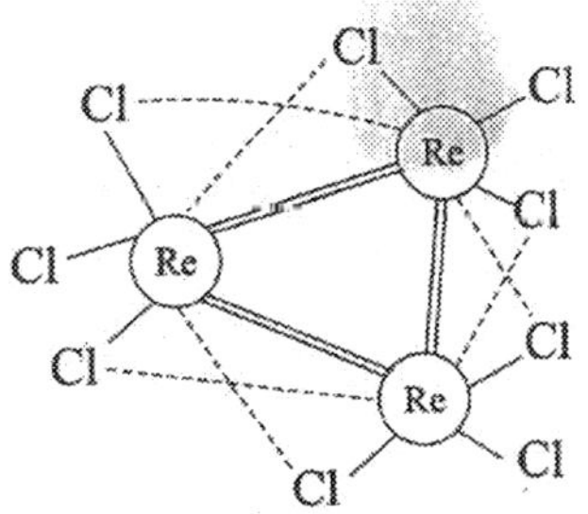

Fig. 6.13 Rhenium (III) trinuclear cluster.

In the triangular array (Re_3 Cl_9] each rhenium atom is coordinated by two more halide ligands above and below the plane defined by the three rhenium atoms. Each Re atom is double bounded to other Re atoms. In the solid state, Re_3Cl_9 is the basic building unit and such units form polymeric structure by bridging Cl atoms.

Tetranuclear Clusters

Tetranuclear carbonyl clusters are $Mo_4(CO)_{16}$, $W_4(CO)_{16}$ and $W_4(OR)_{16}$, with W–W single bonds but the four membered metal ring is not square.

Hexanuclear Clusters

$[Mo_6Cl_8]Cl_4$, is an example of hexanuclear cluster. The octahedrally placed Mo atoms has 4 bonds with adjacent Mo atoms and has 4 coordinate bonds with Cl ligands. There is cubic arrangement of Cl ligands around the Mo atoms.

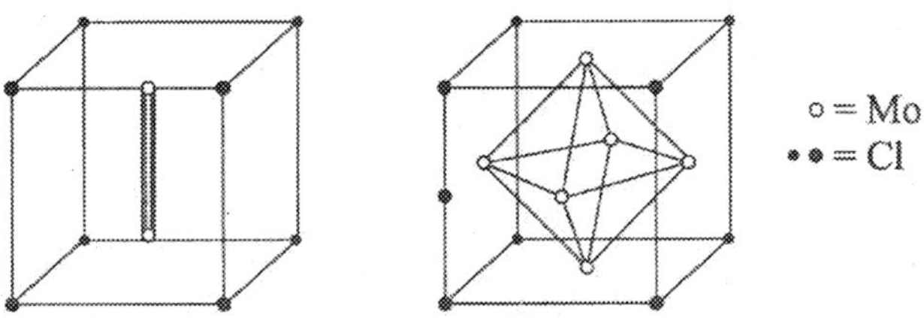

(a) Mo ≡ Mo (b) Mo — Mo

Fig. 6.14 Structure of hexanuclear $[Mo_6\ Cl_8]^{-4}$ complex.

Niobilum, tantalum transition metals also form hexanuclear clusters.

Infinite Metal Chains

In low oxidation state, halides of scandium, yttrium and zirconium are found to have infinite metal bonded chains, e.g. ZrCl contains double metal atom layers alternating with double chlorine layers.

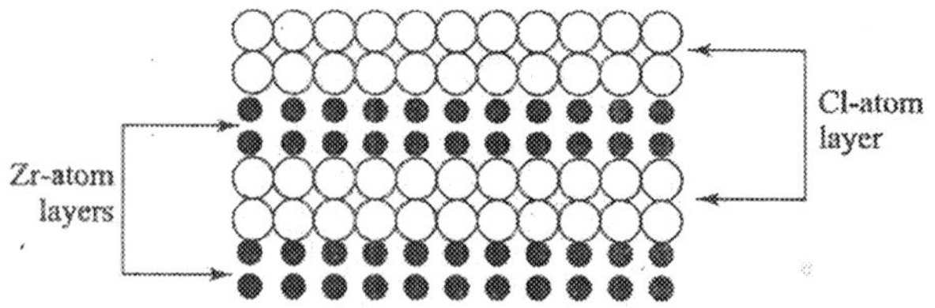

Fig. 6.15 ZrCl containing infinite metal-metal chains.

6.10 METAL OXIDE CATALYSTS

- Partial oxidation reactions are catalysed by transition-metal oxides, enzymes, for example, processes for converting hydrocarbons into valuable products.
- A catalyst is designed to provide a limited amount of oxygen to a reactant for partial oxidation process, allowing formation of desired product but restricting further oxidation.
- The catalysts used for the oxidation are the complex solid oxides, the surfaces of which donate oxygen to the adsorbed organic reactant. It is desired to carry out reaction under particular temperature conditions to avoid unwanted by-products in a carefully designed reactor.

  Catalytic Reactions

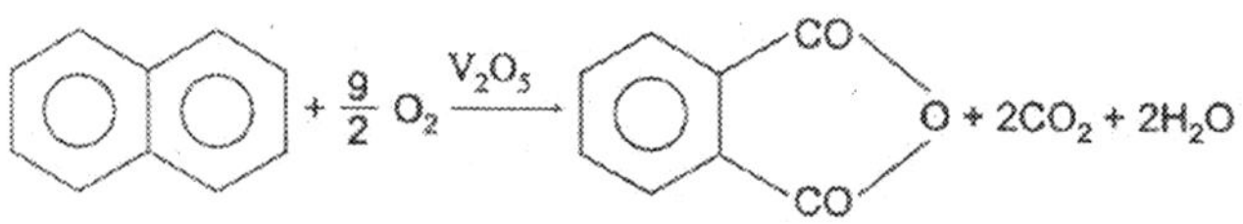

$$CH_3OH + \tfrac{1}{2}O_2 + \xrightarrow{\text{Fe,Mo – oxide}} HCHO + H_2O$$

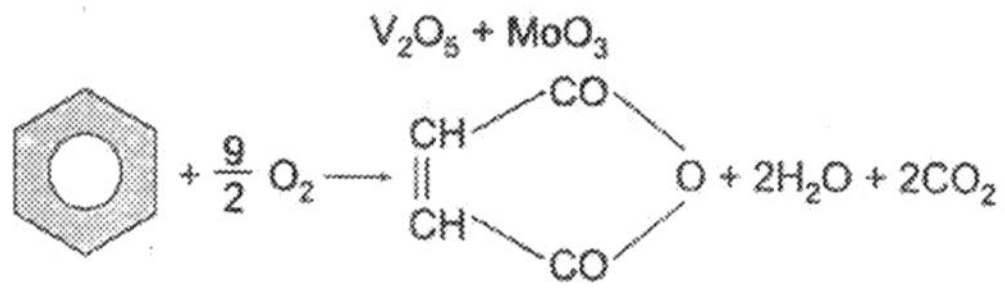

Mechanism

The catalysts for oxidation have the following features:

- The catalyst is selective in particular reaction.
- Binary mixtures of metal oxides are found better than the individual oxide.
- Catalytic oxidation reactions involve single electron transfer via intermediate formations.

- Oxygen atom of adsorbed molecule can form bond with the vacant anion position on catalyst surface or the H can form $\overline{O}H$ on the surface of metal-oxide catalyst, as intermediates. For example,

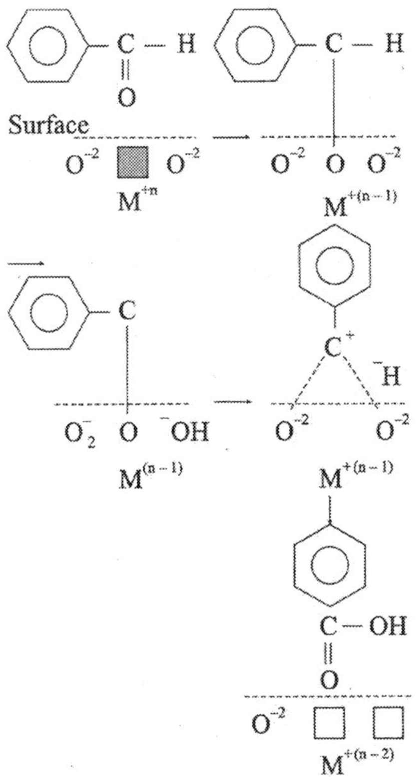

Oxidation of benzaldehyde
to benzoic acid (□ = Vacancy of anion)

Structure of Oxide Surfaces

- A metal ion on surface will have strong adsorption rate and is attracted from all sides by oxide ions, inside the crystal but the metal ion at the surface is not having such balanced forces of attraction. Similarly, the oxide ions on the surface have unbalanced forces of attraction and the adsorption nature thereby.

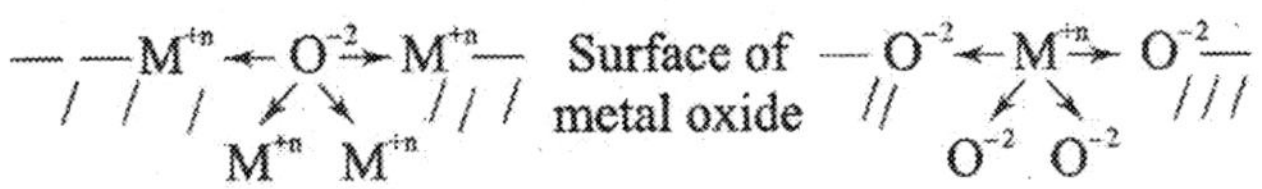

Fig. 6.16 Imbalanced forces of attraction on the surface of an oxide catalyst.

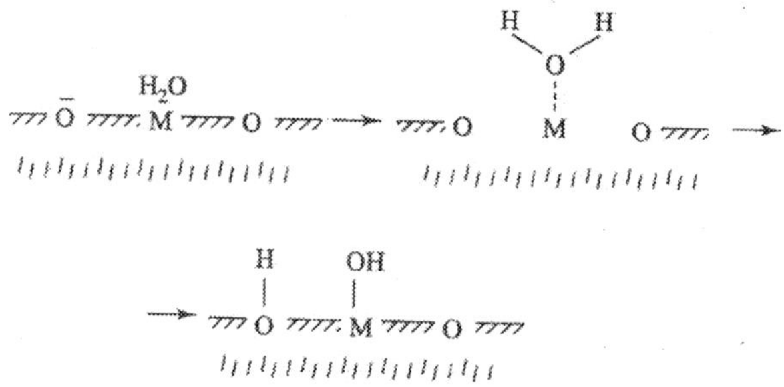

Fig. 6.17 Surface hydration of a oxide catalyst.

- Ligands can get bonded to the metal ions on the surface through surface hydration which can lead to the subsequent reactions. The surface anions vacancies coverage by water, reduces the interaction of alkenes on oxide surface. But water is essential in oxidation processes proceeding through carbocations.

Introduction of oxygen

In the oxidation of hydrocarbon, hydrocarbon is oxidised and oxide is reduced. The reduced oxide reacts with O_2 to give the initial state of catalyst.

$$MO_n \longrightarrow MO_{(n-1)} + \frac{1}{2} O_2$$

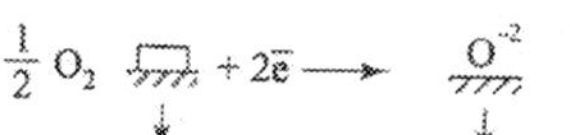

Fig. 6.18 Action of oxidation catalyst.

6.11 PILLARED CLAYS

Natural clays are the metal aluminosilicates. The ions like K^+, Ca^{+2}, Mg^{+2}, Na^+, etc. neutralises the negative charge on the aluminosilicate anions.

- Pillared clays are the chemically modified natural clays or synthetic in which ions like Al^{+3}, Zr^{+2}, Ti^{+4}, H_2O, etc. are sandwiched between the clay anions layers.
- Water between the layered clays can be easily evacuated by heating such clay, which is useful as packing material and potting soil conditioner. e.g. vermiculite.

Types of Pillared Clays

- **Inorganic pillared clays:** Contains inorganic cation sandwiched between the clay anion. e.g. Al- Pil-clay, chromia-Pil- clay, Al-Cu Pil-clay, Ti-Pil-clay, etc.
- **Organic pillared clays:** These contain organic cation sandwiched between clay anion. The organic cation can be carbocations or polymer cations.

Preparation of Pillared Clays

The pillared clays are prepared most efficiently and economically by chemical modification of natural clays. Some of them can be synthesised. Larger sized cations are introduced in place of smaller sized cations like Na^+, K^+, Li^+, H_3O^+, etc. between the clay anions, through cation exchange. Low temperature calcinations of a natural clay with $ZrO/Al_2O_3/TiO_2$, etc. results in chemical bonding of the metal ions with the aluminosilicate ion-layers.

Properties of Pillared Clays

Most of the new pillared clays are designed on nanoscale. Important characteristics of pillard clays are:

Pore size

The pillared clays with larger pore size can act as molecular sieves. Nanosized porous pillared clays are also used specifically. The interlayer and inter-pillar gallery spacing provides large surface area and pore voids.

Cation exchange

Pillared clays possess the property of cation exchange from aqueous solutions. A smaller cation in the pillared clay can be replaced by a larger cation.

Catalytic property

An important property of most of the pillared clays, is their highly selective catalytic nature for particular type of reaction.

Adsorbants

Nanosized pillared clays, act as good adsorbants for acidic and basic compounds, gases, organic dyes (decolorisation) in many reactions because of their larger surface area.

Applications

- They are used for cracking, cyclisation, reduction of NO_x, oxidation of hydrocarbons, ring opening and aromatisation.

- Cationic polymer chains are sandwiched between clay anions, to obtain fire proof and strong composites.
- For separation of different sized molecules or ions in mixture, as a molecular sieve.
- For exchange of cation in water or solution as cation exchangers.
- For removing coloured organic matter from solutions, wastes by adsorption.

6.12 CARBON MEMBRANES

Introduction

Carbon nanotubes may consist of cylinders of graphite sheets of single walled nanotubes (SWNT) or multiwalled nanotubes (MWNT). The nanotubes may be of three types, i.e. zigzag, chiral and armchair, depending upon the arrangement of carbon atoms.

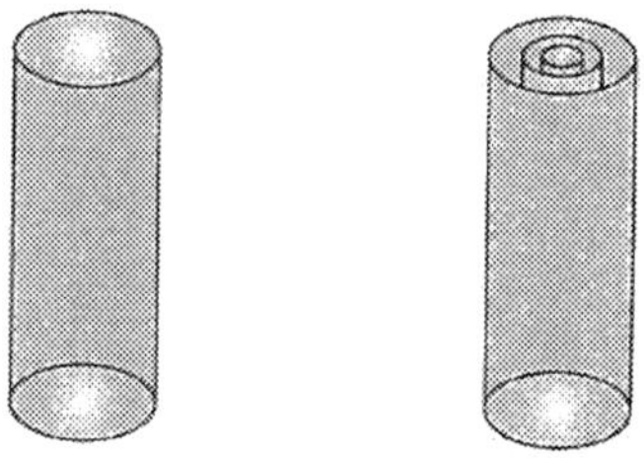

(a) Single walled nanotube (SWNT) (b) Multiwalled nanotube (MWNT)

Fig. 6.19

Membrane Carbon

Carbon nanotubes are hollow and act as pores in the membrane. A carbon membrane consists of a layer of carbon nanbotubes on silicon chip. The supersmooth inner side of the nanotubes can allow free flow of gases and liquids while the larger molecules are blocked by pores.

The membrane is prepared by arranging carbon nonotubes on silicon chip (to support) in single alignment direction. The pores are so small that six water molecules can fit across the diameter of the tube.

Application of Carbon Membranes

Desalination of water is usually done by using reverse osmosis technique under high pressure with less permeable membrane. The water flows easily through the carbon nanotube pores, blocking the salt ions. The cost of desalination is higher

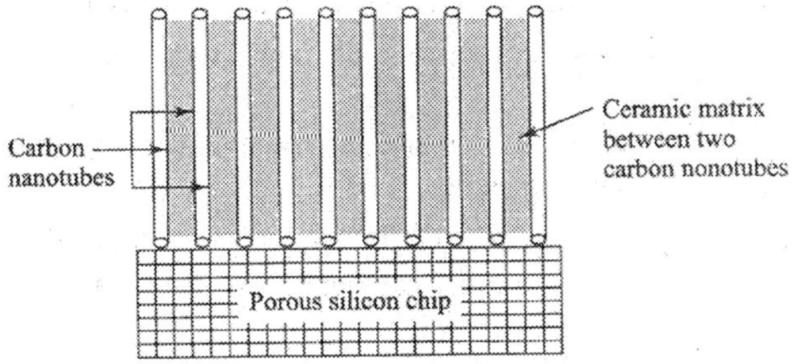

Fig. 6.20 Carbon membrane.

than the other traditional methods. The high gas permeability of carbon membranes is useful for sepearation of industrial hydrocarbon gases. Because of nanometre size and smooth inner surfaces of the nanotubes, carbon membranes are useful in molecular transport and nanofluidics. The membranes help for separation of molecules of different size in a gaseous or liquid mixture. Carbon membranes are designed for adsorption-desorption characteristics study.

QUESTIONS

- What is a catalyst? State the types of catalysts? What are the characteristics of a catalyst?
- Define a promotor and a poison to catalyst.
- Explain theory of unstable intermediate compound formation with its limitations.
- Explain the adsorption theory for heterogeneous catalysis.
- How are the selectivity, activity or efficiency action of promotors, action of catalytic poisons explained on the basis of adsorption theory?
- Define and explain "activation energy".
- Explain the importance of a catalyst in high activation energy reactions?
- Explain how rate constant of a reaction and temperature are related to each other.
- State the importance of catalysts in industrial chemical process.
- What is a zeolite catalyst? Give the important properties of zeolites.
- Describe the adsorption and catalytic properties of zeolites.
- What is a metal complex? Explain the general structures of complexes.
- Explain the role of metal complex catalyst with the help of any two examples.
- Define a metal cluster with the structure of any two binuclear clusters.
- Describe the structure of trinuclear and hexanuclear clusters.

- Explain the role of metal oxide catalyst in organic reactions.
- Explain the mechanism of oxidation by metal oxide catalyst with any one example.
- Explain how oxygen is introduced by metal oxide catalyst during oxidation of organic molecule.
- What is a pillared clay? State the important properties of pillared clays.
- State the physico-chemical characters of pillared clays, with its applications.
- What is carbon membrane? How is it prepared? State its applications.

MUMBAI UNIVERSITY QUESTIONS

May 2008

- What is a catalyst? State types of catalysts. (3)
- Explain adsorption theory of heterogeneous catalysis. (5)
- Explain types of catalysis with suitable examples. (5)
- What is a pillared clay? State important properties of pillared clays. (5)

PRACTICALS

Experiment NO. 1
ESTIMATION OF Zn VOLUMETRICALLY

Title

Estimation of Zn

Aim

To estimate Zinc present in a given sample of brass.

Apparatus

Burette, pipette, conical flask, burette stand, beaker.

Chemicals

0.01 M. EDTA solution, 0.01 M $ZnSO_4$ solution, buffer solution of pH = 10, dil HCl, NH_4OH, Eriochrome Black T indicator.

Theory

Brass is essentially an alloy of copper and zinc with smaller amounts of lead, iron and other elements. To estimate zinc, copper is removed as CuS, lead is removed as $PbSO_4$ and iron is removed as $Fe(OH)_3$. The solution free from all other elements is used to estimate zinc volumentrically.

Procedure

PART – (I) Standardisation of EDTA solution

- Pipette out 25 ml of 0.01 M $ZnSO_4$ solution in a 250 ml conical flask.
- Add about 1/2 test tube buffer solution and 3-4 drops of EBT indicator.

- Titrate against 0.01 M EDTA solution from the burette till it becomes wine red to blue.
- Calculate exact normality of EDTA solution.

Procedure

PART – (II) Opening of the alloy and estimation of zinc

- Weigh accurately about 0.3 gm of given brass. Transfer it in a 250 ml beaker.
- Add about two test tube of dilute HNO_3 to it.
- Warm it till the alloy dissolves completely.
- Evaporate the solution nearly to dryness. Cool the above solution and extract with distilled water and ½test tube of dill HCl.
- Warm the solution again pass H_2S gas to it, till precipitation of CuS is completed.
- Filter this solution by using ordinary filter paper. Wash the residue with hot distilled water.
- Collect filtrate and washing in 250 ml beaker. Again pass H_2S gas, check the precipitation of CuS if ppt. occurs again the process is to be repeated.
- Boil of H_2S completely from the filtrate.
- Cool the above solution to room temperature. Add 1-2 gm of solid NH_4Cl and then add 1:1 NH_3 dropwise to the solution to precipitate $Fe(OH)_3$ and filter it. Wash the precipitate once with hot distilled water. Collect the washing along with the filtrate.
- Evaporate the filtrate almost to dryness. Extract residue with distilled water and transfer to 250 ml standard measuring flask.
- Dilute the solution with distilled water upto the mark and shake the flask well for uniform concentration.
- Pipette out 25 ml of it in a 250 ml conical flask and add about two test tubes of diluted water.
- Add ½ test tube of buffer solution and 3-4 drops of EBT indicator.
- Shake well and titrate solution against 0.01 M EDTA solution from the burette. End point will be from wine red to blue colour.

Observations

Part – (i) Standardisation of EDTA solution

Solution in Burette – 0.01 M EDTA solution. (Appoximate)

Solution in Pipette – 25 ml of 0.01 M $ZnSO_4$ solution

Indicator – Eriochrome Black T.

End point – Wine red to blue.

Pilot reading – .... ml to .... ml

Chemical Reaction

$$Zn + 2Na_2C_{10}H_{14}O_8N_2 \rightarrow ZnC_{10}H_{14}O_8N_2 + 2Na^+$$

Observation Table

Sr. No.	*Initial B.R.* (ml)	*Final B.R.* (ml)	*Difference B.R.* (ml)	*C B.R.* (ml)	
1					
2					
3					= 'x' ml

Calculations

$M_1V_1 = M_2V_2$ (M_1V_1 = EDTA, $M_2V_2 = ZnSO_4$)

Therefore, $M_1 = M_2V_2/V_1$

Therefore, $M_1 = 0.01 \times 25/X$

Therefore, M_1 = AM. (A – Molarity of EDTA)

Observation

Part – (II) Estimation of Zinc

Solution in Burette – 0.01 M EDTA solution. (AM)

Solution in Pipette – 25 ml of brass samples solution

Indicator – Eriochrome Black T.

End point – Wine red to blue.

Pilot reading – .... ml to .... ml

Chemical reaction

$$Zn + 2Na_2C_{10}H_{14}O_8N_2 \rightarrow ZnC_{10}H_{14}O_8N_2 + 2Na^+$$

Observation Table

Sr. No.	Initial B.R. (ml)	Final B.R. (ml)	Difference B.R. (ml)	C B.R. (ml)	
1					
2					
3					= 'y' ml

Calculations

1000 ml of 1 M EDTA = 65.38 gm of Zn

25 ml of brass sample solution required 'Y' ml of AM. EDTA solution.

25 ml of brass sample solution = 65.38 × Y × A/1000 = B gm of Zn

250 ml of brass sample solution = B × 10 = C gm of Zn

% of Zinc = C × 100/W (W = Weight of brass sample)

Therefore, % of Zn in given sample of brass = … %.

Result

1. Molarity of EDTA solution = … M.
2. Percentage of Zinc in the given sample of Brass = … %.

Experiment NO. 2
ESTIMATION NICKEL GRAVIMETRICALLY AS DIMETHYL GLYOXIME

Aim

Estimate the amount of Nickel present in the given Nickel Sulphate solution

Requirements

DMG, nickel sulphate solution, ammonia, glass beaker, funnel, oven, weighing balance, etc.

Theory

Nickel ion in ammonical medium reacts with an organic reagent Dimethyl glyoxime and forms a rosy red chelate compound, which is dried at 110°C. Dimethyl glyoxime chelating agent, has great importance in analytical chemistry. Chelate compounds are often formed quantitatively with very small or even with traces of ion present, when specific reagents are used under controlled conditions and such compounds are quite stable.

Procedure

- Dilute the given solution to 250 ml in a standard measuring flask with distilled water.
- Pipette out 50 ml of it in 250 ml beaker and add 1 to 2 test tubes of H_2O.
- Add 30-35 ml of 1% dimethy glyoxime reagent to the solution and stir it well.

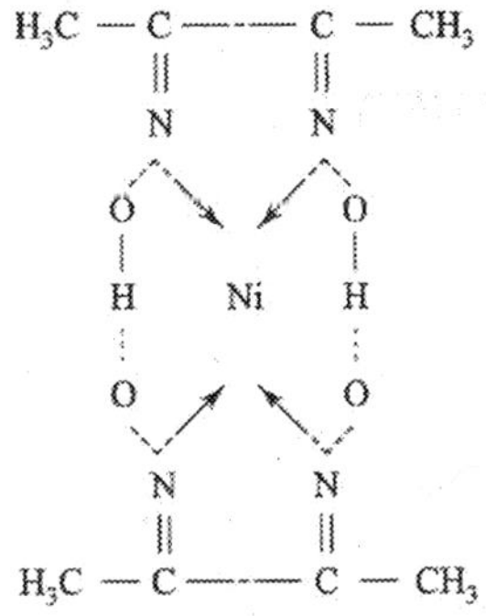

Fig. Ni-dimethyl glyoxime.

- Add 6 N NH_4OH dropwise till the precipitation of Ni^{+2} as Ni d.m.g. is complete. Smell of NH_3 should be distinct to ensure alkaline medium.
- Digest the precipitate on a waterbath for 20-30 minutes.
- Test the supernatant solution for the complete precipitation with 1-2 drops of the reagent.
- Filter the precipitate through counterpoised filter papers No.41.
- Wash the precipitate with distilled water, till the filtrate is free from chloride and sulphate ions. Finally, wash it with alcoholic water.
- Dry the precipitate in an oven at 110°C and weigh it as Ni-dimethylglyoxime. (Ni dmg)

Reactions

$$NiSO_4 + (C_4H_3N_2O_2)_2 \rightarrow Ni(C_3H_7N_2O_2)_2 + H_2SO_4$$

Calculations

Let the weight of residue (Ni dmg) be 'x' gm.

Now, 50 ml of dilute solution = 'x'gm of Ni.d.m.g.

∴ 250 ml of dilute solution = 5 'x' gm

Ni d.m.g. = Ni

288.71 gm Ni.d.m.g. = 58.71 gm Nikel

$$\therefore 5\text{ 'x' gm Ni d.m.g.} = \frac{58.71 \times 5\text{'x'}}{288.71}$$

$$= \ldots\text{gm of Nickel}$$

Results

(1) 50 ml of dilute solution gave ...gm of Ni. d.m.g.

(2) Amount of nickel present ...gm.

Experiment NO. 3
SYNTHESIS OF BIODIESEL

Title

Biodiesel synthesis

Aim

Synthesis of biodiesel and its characterisation

Apparatus

Stirrer, beaker, weighing balance, etc.

Chemicals

Vegetable oil, KOH, methanol

Theory

Vegetable oil and animal fats are basically triglycerides. They are fatty acid esters of a single alcohol glycerol. Vegetable oils and animal fats have very high viscosity and they cannot be used directly as fuel. By transesterification reaction, triglycerides can be converted into their methyl or ethyl esters, having less viscosity w.r.t. oils and fats. These methyl or ethyl esters can be used as biodiesel. Biodiesel is a renewable source of energy, having low pollution potential than conventional diesel or a blend of diesel and biodiesel is used.

The transesterification given below can be carried out either by using acid or base catalyst.

Procedure

Biodiesel is less expensive as compared to conventional diesel can be synthesised as:

Moisture free oil or fat (with less than 5% free acid) and dry methyl or ethyl alcohol are taken in 1:3:5 molar ratio, in a three-neck round bottom flask (in two side necks, thermometer pocket and condenser are fitted and through middle neck, overhead stirrer is introduced).

If acid catalyst is used, 1:30 molar ratio of sulphuric acid and oil are taken. Reaction requires 69 hours.

For base catalysis, oil and base (NaOH or KOH) are used in 1:1 molar ratio. Stirring with six hundred revolutions per minute for two hour is required.

For reaction to take place, heating at 60°C is required for methanol and 70°C for ethanol. Completion of reaction is confirmed by taking TLC.

Result

.... % yield of methyl or ethyl ester.

Characterisation may be done by measuring the boiling point and viscosity of vegetable oil and methyl or ethyl ester. Boiling point of oil may be taken by using silicon oil bath.

Sr. No.	Description	Boiling point in °C	Viscosity in poise
1	Vegetable oil		
2	Methyl or ethyl ester		

Reaction

$$\begin{array}{lcll} CH_2 - OOC - R_1 & & R_1 - COOCH_3 & CH_2OH \\ \vert & \xrightarrow[\text{Heat}]{CH_3ONa} & & \vert \\ CH - OOC - R_2 & & R_2 - COOCH_3 \; + & CHOH \\ \vert & & & \vert \\ CH_2 - OOC - R_2 & & R_3 - COOCH_3 & CH_2OH \\ \text{Veg·oil or fat} & & \text{Biodiesel} & \text{Glycerol} \end{array}$$

Experiment NO. 4
DETERMINATION OF MOISTURE CONTENT OF THE COAL

Title

Moisture content

Aim

Estimation of moisture content in a given sample of coal

Apparatus

Oven, silica crucible, desiccators, etc.

Chemicals

Powdered coal sample

Theory

In order to ascertain the quality of coal, proximate and ultimate analysis of coal is important. Proximate analysis is determination of moisture content, volatile matter, ash content and fixed carbon content. This gives information about the practical utility of coal and helps in grading of coal.

Moisture content is the loss in weight of coal when heated in crucible at 110°C for one hour. During burning of coal, moisture evaporates and is removed as steam. Some amount of heat is required for this, which is taken during the burning of fuel. Calorific value of fuel is reduced due to the presence of moisture in fuel.

Procedure

Take accurately weighed quantity of powdered coal sample in previously weighed silica crucible. Heat the sample in an electric oven at about 100 to 110°C for one hour. Take out the crucible, cool it in desiccator and weigh it. Repeat the procedure of heating, cooling and weighing of the crucible till constant weight is obtained. Note down the constant weight.

Observation Table

Sr. No.	Description	Value
1	Weight of empty crucible	W_1g
2	Weight of crucible and coal	W_2g
	Weight of coal	$(W_2 - W_1)$ g
3	Weight of crucible and coal after heating	W_3g
4	Weight of coal after heating	$(W_3 - W_1)$ g

Calculations

Loss in weight due to heating = moisture content of sample

$$= (W_2 - W_1)\ (W_3 - W_1)\ g$$

$$= (W_2 - W_3)\ g$$

% moisture = loss in weight/weight of coal × 100

$$= (W_2 - W_1)/(W_3 - W_1) \times 100$$

Results

Percentage of moisture in given sample of coal = ... %

Experiment NO. 5
ESTIMATION OF NICKEL (VOLUMETRICALLY)

AIM

To determine the percentage of nickel in stainless steel sample.

Requirements

Nickel, 0.05 M EDTA, EBT, Murexide, buffer solution, etc.

Principle

When a Nickel solution is titrated against EDTA at pH –10 using Murexide as indicator, the free Ni-ion are first complexed with the indicator. Then the Ni-ions from the indicator complex are displaced to form the more stable Ni-EDTA complex liberating the free form of the indicator. It is indicated by the colour change from yellow to blue violet.

Procedure

Part-I

A known weight of steel sample (0.1 to 0.2 gm.) is taken into a 250 ml beaker and dissolved in 10 ml. of conc. HCl. The sample is oxidised by the dropwise addition of conc. HNO_3 and digested by keeping it over the hot plate. The solution is then transferred to 100 ml. Standard measuring flask and made up to the mark.

Part-I Standardisation of EDTA solution

25 ml of standard $ZnCl_2$ is pipetted into a clean conical flask 5 ml of buffer and 2-3 drops of EBT are added and titrated against EDTA solution taken in the burette. The end point is change of colour from wine red to blue. The titration is repeated for concordant values from this titre values the strength of EDTA is calculated.

Part-II Estimation of Nickel

25 ml of made up (or diluted) solution is pipetted into a clean conical flask. The excess acid is neutralized by adding buffer to this murexide is added. If the solution turns pale yellow, add few more drops of buffer till a deep yellow color is formed. This is then titrated against EDTA from the burette. The end point is violet.

(If the solution turns pale yellow, buffer is added in drops during the titration.)

Observation

Part-1 Standardisation of EDTA solution

Solution in Burette – 0.05 M (approximate)

Solution in Pipette – 25 ml of 0.01 M. $ZnSO_4$ solution
Indicator – Eriochrome Black T.
End point – Wine red to blue.
Pilot reading – .... ml to .... ml.

Observation Table

Reading	*Final* (ml)	*Initial* (ml)	*Difference* (ml)	*CRB*(ml)
1				
2				
3				

= 'x' ml

Calculation

$M_1V_1 = M_2V_2$ (M_1V_1 = EDTA, $M_2V_2 = ZnSO_4$(7420))

Therefore, $M_1 = M_2V_2 = \frac{M_2V_2}{V_1}$

Therefore, $M_1 = \frac{0.05 \times 25}{X}$

Therefore, M_1 = AM. (A = Molarity of EDTA)

Part-11 Standardisation of EDTA solution

Solution in Burette – 0.05 M EDTA (approximate)
Solution in Pipette – 25 ml of steel sample solution
Indicator – Murexide.
End point – Pale yellow to violet.
Pilot reading – ....ml to ....ml.

Observation Table

Reading	*Final* (ml)	*Initial* (ml)	*Difference* (ml)	*CBR* (ml)
1				
2				
3				

= 'y' ml

Calculation

1000 ml of 1M EDTA ≡ 58.71 of Ni.
25 ml of brass sample solution required Y ml of AM.EDTA solution.

$$25 \text{ ml of brass sample solution} = \frac{58.71 \times Y \times A}{1000}$$

$$= B \text{ gm of Ni.}$$

250 ml of steel sample solution = B × 10 = C gms of Ni

$$\% \text{ of Ni} = \frac{C \times 100}{W} \quad (W = \text{Weight of steel sample})$$

$$= \ldots\ldots.\%$$

Therefore, % of Ni in given sample of steel......%

Result

1. Molarity of EDTA solution = ... M
2. Amount of nickel present in the given sample = ...%

Experiment No. 6
ESTIMATION OF COPPER IN BRASS

AIM

To estimate the amount of copper in brass iodometrically.

Requirements

0.1N.(Approx) Sodium Thiosulphate solution, 10% Potassium Iodide solution, conc. HCl, 0.1 N Potassium Dichromate solution, Acetic Acid Glacial, Solid Sodium Carbonate, Starch Indicator, etc.

THEORY

The chief components of brass are copper and zinc. But small quantities of tin and other metals are also usually present. The copper content varies from 60 to 90 %. Copper present in brass is converted in to cuprous iodide. The cuprous iodide solution is titrated with standard sodium thiosulphate solution using starch as an indicator.

Procedure

Part- I Standardisation of Sodium Thiyosulphate solution

1. Pipette out 10 ml of 0.1 N potassium dichromate solution in a 250 ml flask. Add about 3 ml of conc. HCl and 10 ml of 10% potassium iodide solution to it. Shake it well.

2. Titrate the lilberated iodine against 0.1 N sodium thiosulphate solution from the burette using freshly prepared starch indicator till it becomes blue to pale green.

Part-II Estimation of Copper in Brass

1. Dissolve given sample (accurately weighed) in conc. HCl in 250 ml beaker (This treatment ppts Tin as metastanic acid but copper and zinc are dissolved).
2. Metastanic acid is filtered out and the washings are collected in a 250 ml beaker.
3. Copper nitrate solution is converted into copper sulphate by evaporating the solution with conc. sulphuric acid nearly to dryness extracted with water and diluted to 250 ml in standard measuring flask.
4. Pipette out 25 ml from standard measuring flask of the diluted solution in a 250 ml conical flask.
5. Add pinch of solid of sodium carbonate till it gets slightly turbid.
6. Dissolved the turbidity by adding 2 N acetic acid drop wise till it becomes clear.
7. Add 10 ml of 10% potassium iodide solution and shake it well. Titrate it against sodium thiosulphate from the burette till it becomes blue to colourless.

Observations

Part-1 Standardisation of Sodium Thiyosulphate solution

Solution in burette – 0.1 N sodium thiosulphate solution.

Solution in pipette – 10 ml. of 0.1 N potassium dichromatic solution.

Indicator – Starch.

End point – Green to colourless.

Pilot reading – ..... ml to ..... ml.

Reactions

1. $K_2Cr_2O_7 + 14HCl \rightarrow 2KCl + 2CrCl_3 + 3Cl_2 - 7H_2O$.
2. $Cl_2 + 2KI \rightarrow 2KCl + I_2$.
3. $I_2 + 2Na_2S_2O_3 \rightarrow 2NaI + Na_2S_4O_6$.

Observation Table

Reading	*Final* (ml)	*Initial* (ml)	*Difference* (ml)	*CRB* (ml)
1				
2				
3				

= 'x' ml

Calculation

10 ml of 0.1 N potassium dichromate solution required 'x' ml of sodium thiosulphate.

$N_1V_1 = N_2V_2$ (N_1V_1 = Sodium thiosulphate, N_2V_2 = potassium dichromate)

Therefore, $N_1 = \dfrac{N_2V_2}{V_1}$

Therefore, $N_1 = \dfrac{0.1 \times 25}{X}$

Therefore, $N_1 = AN$ (A – Normality of sodium thiosulphate)

Part-II Estimation of Copper

Solution in burette – 0.1N. sodium thiosulphate.

Solution in pipette – 25 ml of brass sample solution

Indicator – starch.

End point – green to colourless.

Pilot reading – ...ml to ...ml.

Reaction

1. $2CuSO_4 + 4KI \rightarrow Cu_2I_2 + 2K_2SO_4 + I_2$.
2. $I_2 + 2Na_2\,S_2O_3 \rightarrow 2NaI + Na_2S_4O_6$.

Observation Table

Reading	*Final* (ml)	*Initial* (ml)	*Difference* (ml)	*CBR* (ml)
1				
2				
3				

= 'y' ml

Calculation

1000 ml of sodium thiosulphate = 63.54 gms. of Cu.

25 ml. of brass sample solution required Y ml of A. N. sodium thiosulphate solution.

25 ml. of brass sample solution $= \dfrac{63.54 \times Y \times A}{1000}$

= B gms. of Cu

250 ml of brass sample solution = B × 10 = C gm

$$\% \text{ of Cu} = \frac{C \times 100}{W} \quad \text{(W–Weight of brass sample)}$$

Therefore = ...% of Cu in given sample of Brass = ... %

Results

1. Normality of sodium thiosulphate solution = ... N.
2. Percentage of copper in the given sample of brass = ... %

Experiment NO. 7
ESTIMATION OF TIN VOLUMATRICALLY

AIM

To determine amount of Tin present in stannous chloride solution obtained from solder alloy.

Requirements

0.05 N (Approx) iodine solution, standard 0.05 N sodium thiosulphate solution($Na_2S_2O_3$) solution, freshly prepared starch indicator.

Theory

Solder alloy consists of about 3 parts (55-69 %) of Pb and 2 parts about (35-40 %) of Sn in additional to traces of antimony.

The alloy is dissolved in conc. HNO_3, when Sn and traces of Sb remains insoluble while Pb goes in solution. Solution is filtered through paper no. 42. Residue on washing is dissolved in conc. HCl and diluted solution is titrated against std. iodine solution.

Procedure

Part-I Standardisation of Iodine solution

Pipette out 10 ml of std. 0.05 N Na_2 S_2O_3 solution in a 250 ml conical flask. Add about 10 ml of dil. HCl (one test tube) to it. Then adding about two test tubes of distilled water and 1-2 ml of starch indicator titrate against iodine solution from burette till it becomes colourless to blue.

Part-II Estimation of TIN

Pipette out 25 ml of given solution of stannous chloride in a 250 ml conical flask. Then add about 1-2 ml of freshly prepared starch indicator shake well. Titrate it with iodine solution from the burette till faint blur colour persists.

Observations

Part-1 Standardisation of Iodine solution

Solution in burette – 0.05 N. iodine solution.

Solution in pipette – 10 ml. of 0.05 N $Na_2S_2O_3$ solution

Indicator – Starch.

End point – Colourless to blue

Pilot reading – ...ml to ...ml.

Reactions

1 $I_2 + 2Na_2S_2O_3 \rightarrow 2NaI + Na_2S_4O_6$.

Observation Table

Reading	*Final* (ml)	*Initial* (ml)	*Difference* (ml)	*CBR* (ml)
1				
2				
3				= 'x' ml

Calculation

10 ml of 0.05 N sodium thiosulphate solution required required X ml of Iodine.

$N_1V_1 = N_2V_2$ (N_1V_1 Iodine – N_2V_2 – sodium thiyosulphate)

$$\text{Therefore, } N_1 = \frac{N_2V_2}{V_1}$$

$$\text{Therefore, } N_1 = \frac{0.5 \times 25}{X}$$

Therefore, N_1 = AN (A – normality of iodine)

Part-II Estimation of TIN

Solution in burette – AN. iodine solution

Solution in pipette – 25 ml. of stannous chloride sample solution

Indicator – Starch

End point – Colourless to blue

Pilot reading – ...ml to ...ml.

Reaction

1. $SnCl_2 + 2HCl + I_2 \rightarrow SnCl_4 + 2HI$

Observation Table

Reading	*Final* (ml)	*Initial* (ml)	*Difference* (ml)	*CBR* (ml)
1				
2				
3				

= 'y' ml

Calculation

1000 ml of 1N Iodine = 59.35gm of Sn.

25 ml of stannous chloride sample solution required Y ml of A N. iodine solution.

$$25 \text{ ml. of brass sample solution} = \frac{59.35 \times Y \times A}{1000}$$

$$= B \text{ gms of Sn.}$$

250 ml of stannous chloride sample solution = B × 10 = C gms

$$\% \text{ of Alloy} = \frac{C \times 100}{W} \quad (W - \text{Weight of brass sample})$$

Therefore, % of tin in solder alloy = ...%

Experiment NO. 8
DETERMINATION OF CALORIFIC VALUE

Aim

To determine the calorific value of a fuel sample by using a bomb calorimeter.

Requirements

Bomb calorimeter, fuel, sample, benzoic acid, glassware, etc.

Theory

Calorific value of a fuel is the amount of heat liberated by the complete combustion of a unit mass of fuel. The calorific value determines the quality of a fuel. The higher calorific value can be calculated as

$$HCV = \frac{(W + w)((t_2 - t_1) \text{ or } \Delta t + \text{cooling correction}) - (\text{Acid + fuse wire correction + thread corr.})}{m}$$

where w = mass of water taken in calorimeter

W = water equivalent of calorimeter

m = mass of fuel taken in bomb

$(t_2 - t_1) = \Delta t$ = difference of initial and final temperature of calorimeter water

and LVC = HCV – (0.09H × 587)

where H = % of hydrogen in coal

Procedure

Take 0.5 g of given fuel in a dry steel crucible. A fine platinum/nichrome wire, touching the sample fuel is stretched across the electrodes and a cotton thread of known length is tied on fuse wire and inserted in sample inside. Place 10 ml of water inside the bomb and tightly close the bomb cover with lid. Now charge the bomb by O_2 gas applying O_2 cylinder to get the pressure ~25 atmosphere. Now bomb is placed inside the calorimeter. Add known mass of water say 2 litres in calorimeter so that the bomb is submerged in water but terminals remain above the water level. There should be no bubble or leakage in the bomb. Connect the terminals with electricity. Switch on the stirrer and note the initial temperature of water. After burning the fuel, the water should be stirred for half an hour to get maximum temperature. Leave the bomb for one hour to settle down the fumes. Now release the oxygen of bomb and open the bomb, wash the residue of bomb with water and transfer to a beaker. Use this washing to determine the presence of HNO_3 and H_2SO_4.

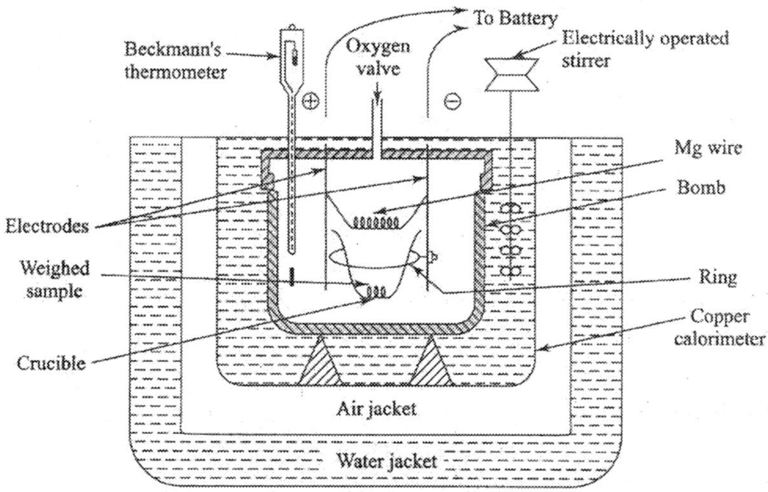

Fig. Bomb calorimeter

Determination of H_2SO_4 and HNO_3: (Acid correction)

Due to the presence of sulphur in fuel, S is oxidised to SO_2 which dissolve in water to give H_2SO_4. This reaction is exothermic hence, a necessary correction against H_2SO_4 formed should be made in calculation of H.C.V.

$$S + O_2 \longrightarrow SO_2$$

For every gram equivalent of H_2SO_4 (49 gm H_2SO_4) 36 kcal of heat is librated.

$$1 \text{ ml of } \frac{N}{10} H_2SO_4 \equiv 49 \times \frac{N}{10} \times \frac{1}{100} \text{ g } H_2SO_4 \cong \frac{36}{10^4} \text{ kcal} = 3.6 \text{ cal.}$$

Similarly, $1 \text{ ml of } \frac{N}{10} HNO_3 \equiv 63 \times \frac{N}{10} \times \frac{1}{1000} \text{ g } HNO_3 \cong 1.429 \text{ cal.}$

Dilute the washing from bomb to 100 ml and boil to remove CO_2.

Now titrate it with $\frac{N}{10}$ $Ba(OH)_2$/NaOH using methyl orange as an indicator.

Add 20 ml $\frac{N}{10}$ Na_2CO_3 solution. Warm and filter the solution. Wash the precipitate and titrate the supernatant with $\frac{N}{10}$ HCl using methyl orange as an indicator.

Let $\quad \frac{N}{10}$ HCl used = A ml

$\frac{N}{10}$ $Ba(OH)_2$ used = B ml

$\therefore \quad \frac{N}{10}$ HNO_3 = (20 – A) ml ⇒ 1.43(20 – A) cal.

$\frac{N}{10}$ H_2SO_4 = (A + B – 20) ml ⇒ 3.6(A + B – 20) cal.

Fuse wire correction: For 1″ of fuse wire = 35 cal/g of nichrome wire.

Cotton thread correction: For 1g of cotton = 4140 cal/g.

Cooling correction: $\quad t_{\text{cooling}} = nV + \left(\frac{V' - V}{t' - t}\right)\left[\Sigma(t) + \frac{t_o + t_n}{2} - nt\right]$

where,

n = No of minutes between firing and maximum temp.

V = Rate of fall of temp. during initial period before firing.

V' = Rate of fall of temp. during final period after maximum temp.

t = Mean temp. during initial period.

t' = Mean temp. during final period.

$\Sigma(t)$ = Mean temp. during initial to final period.

t_o = Temp. at the time of firing.

t_n = Temp. at the time of maximum value.

Observation

S. No.	Initial period		Combustion period		Final period	
	Time	°C	Time	°C	Time	°C
1						
2						
3						
4						
5						
6						

Result

The H.C.V. of given fuel = ...cal/g.

Precautions

1. The amount of sample taken should not be more than 1.0 g.
2. There should not be any leakage in bomb.
3. The fuel should be finally powdered before placing in bomb.
4. The bomb should not be opened just after the firing otherwise some fumes may blow out.
5. All the parts of bomb should be submerged in water except electrode terminals.
6. The water should be stirred all the time.

Experiment NO. 9
GAS ANALYSIS

CO_2 Gas Analysis by Orsat Apparatus

When a fuel is burnt in air, the products of combustion are CO_2, CO, water vapour some unburnt fuel which escapes during combustion. The mixture of gases coming out of combustion chamber is called as flue gas. The proportion of various gases present in this gases will vary depending upon the amount of air used, efficiency of mixing fuel with air, dimension of combustion chamber and efficiency of exhaust system. Generally, a slight excess of air is used to ensure complete combustion and derive the maximum heat output. Orsat apparatus is used for gas analysis.

Preparation of Absorbent Solution

1. **For CO_2 absorption**

   300 gm of KOH pellets dissolved in distilled water and volume made upto a litre. The solution is stored in rubber stoppered bottle and is decanted for use.

2. **Absorbent for Oxygen**

   To 500 ml of KOH solution prepared as above, 80 gm pyrogallol is added and stirred well.

3. **Absorbent for carbon monoxide**

   75 gm of cuprous chloride is mixed with 15 gm of ammonium chloride and 80 ml of water. Concentrated ammonia is added to the mixture till all cuprous chloride dissolves completely. This solution is kept in contact with copper turnings.

Procedure

The inlet tube is attached to U-tube containing dry $CaCl_2$. The stopcocks in all the pipettes are properly greased. The absorbents are transferred to tubes in the following order. The first absorbent pipette is filled with KOH solution.

The second absorbent pipette is filled with alkaline pyrogallol.

The third absorbent pipette is filled with ammoniacal cuprous chloride solution.

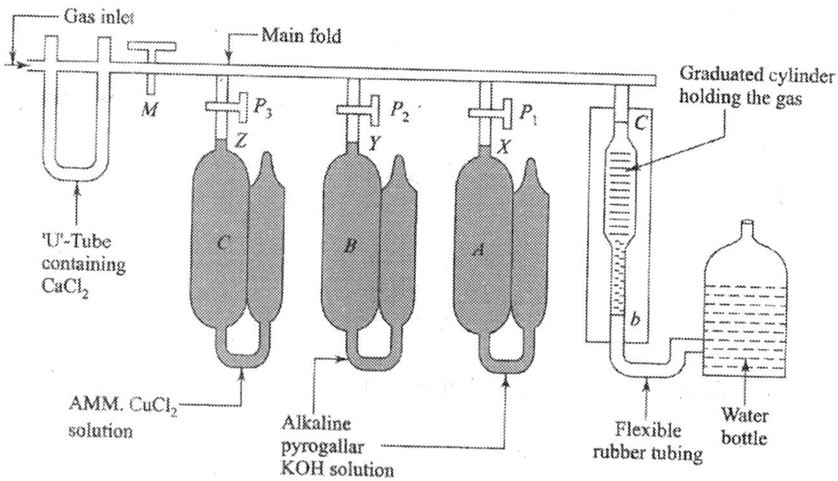

M-Main Three Way Stopcock

P_1, P_2, P_3, stopcocks of absorption tube

X, Y, Z, marks so absorption tubs to keep the liquid elvel.

a, b, marks on gas holder to indicate 0 and 100 ml volume.

Working of Orsat Apparatus

1. After having filled the absorbent in the tubes stopcock *M* is opened. The water bottle with water lift to force the air in gas holder out. At this stage stopcocks P_1, P_2, P_3 will be in closed position. The level is raised so that the water level reaches position "a" and immediately stopcock *M* is closed.
2. When the water bottle is lowered, the position of level in gas holder will not come down if there is no leak anywhere.
3. The level of absorbents in the tubes *A*, *B*, *C* are brought to *X*, *Y*, *Z* by carefully opening the stopcocks P_1, P_2, P_3. Once adjusted they should remain in the same position.
4. Flue gas is now drawn into gas holder by attaching to inlet tube through U-tube containing anhydrous $CaCl_2$ to absorb any moisture and by opening the stopcock *M*.

Water level in the gas holder will rapidly fall down and as it reaches the mark "*b*" the stopcock *M* is closed.

100 ml of flue gas holder will now be successively passed through absorbents as follows:

(1) By lifting the water bottle and slightly opening the stopcock P_1, the flue gas is gently transferred into the absorbent tube A containing KOH. The level of KOH in absorbent tube *A* will go down as the water level in gas holder goes up. As the water bottle is lowered, the KOH level in *A* will go up, absorbing any CO_2 present. This process of lifting and lowering of water bottle is done for 4–5 times and finally the level of KOH in tubes is brought back to X and stopcock closed. The water level in the bottle and that in the gas holder are brought to the same level. It will be observed that level has gone up indicating that part of gas is absorbed. For example, the initial volume of gas taken is 100 ml. After passing through KOH tube, the volume of remaining gas is 9–90 ml. That means 10 ml of CO_2 has been absorbed by KOH tube.

(2) This 90 ml of remaining gm is then passed through absorbent tube *B* containing alkaline pyrogallol which absorbs oxygen by gently opening the stopcock P_2 and rising the water bottle. As water rises in the gasholder, it pushes gas into tube *B*. As water bottle is lowered, the displaced reagent rises absorbing oxygen present in gas. Again this process of rising and lowering of water bottle is done a couple of times to ensure that all the gas is exposed to alkaline pyrogallol. Finally, the bottle is lowered so that level of absorbent is brought back to level *y* in tube *B* and stopcock P_2 closed. The level in bottle and gas holder are once again equalised and volume of gas still remaining is noted down. Suppose it 82 ml. That means 8 ml of oxygen is absorbed by absorbent in *B*.

(3) 82 ml of remaining gas is now passed through the last absorbent by adopting the procedure indicated earlier. Finally, the volume after absorption is noted in gas holder, equalising the levels in bottle and gasholder. Generally, CO content is low the volume may be somewhere 81.0 ml. Hence, the flue gas composition can be expressed as

10% CO_2

8% O_2

1% CO

The experiment can be repeated by taking 100 ml fresh flue gas and passing through the absorbent as before.

Result

Volume of flue gas taken			100 ml
Volume after passing through KOH tube			x ml
Volume after passing through pyrogallol tube			y ml
Volume after passing through amm. $CuCl_2$ solution			z ml
% CO_2 =	(100 – x) ml	=	
% O_2 =	(x – y) ml	=	
% CO =	(y – z) ml	=	

MODEL QUESTION PAPER

Time = 2 hrs | Marks = 75

N.B.:

1. Question No.1 is compulsory.
2. Attempt any four questions from remaining six.
3. Figures to the right indicates full marks.

1. Answer the following (any five) (15)
   (a) Why silver, gold and platinum do not undergo oxidation corrosion?
   (b) How rate of corrosion is influenced by design of the material?
   (c) Give composition properties of German silver.
   (d) Define octane number and cetane number.
   (e) Explain the significance of carbon and hydrogen in a coal.
   (f) Calculate % atom economy in moleic an hydride production.

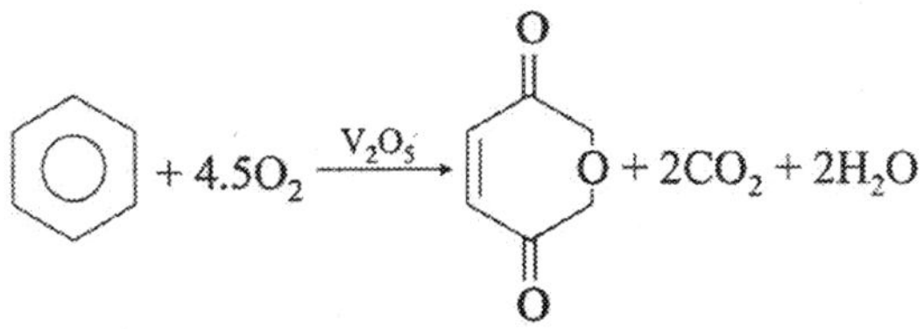

   (g) What are composite materials? Write their applications.
   (h) Define promoter and poison to catalyst.
2. (a) Define corrosion. Explain mechanism of electro-chemical corrosion. (5)
   (b) Explain fraction distillation of crude petroleum. (5)
   (c) What are the applications of powder metallurgy ? (5)
3. (a) (i) Same quantity of electricity is passed through copper sulphate and silver nitrate solution. If 0.259 gm of copper and 0.901 gm of silver are deposited. If the eq. ut of copper is 31.6 find eq. ut. of silver. (3)
   (ii) Give composition and uses of Duralium. (2)
   (b) Explain determination of moisture content and volatile matter present in a coal. (5)

(c) Define green chemistry ? Explain following principles of green chemistry (5)

(i) Non-hazardous chemical synthesis

(ii) Use of catalytic reagents.

4. (a) Explain cathadic protection method for control of corrosion. (5)

(b) Explain fixed bed catalytic cracking. (5)

(c) Calculate the volume of oxygen required for complete combustion of 1 m^3 of a gaseous fuel of following composition by volume.

CH_4 = 5%, CO = 25%, H_2 = 10%, N_2 = 50%, CO_2 = 5%, O_2 = 2%.

5. (a) Explain significance of green chemistry. (5)

(b) Explain sandwich panel composites with suitable examples. (5)

(c) What are the important properties of zeolites. (5)

6. (a) State physico-chemical characters of pillard clays with their applications. (5)

(b) Explain homogenous catalysis. (5)

(c) What are the characteristics of matrix and particulate phase. (5)

7. Write notes on (any three) (15)

(a) Tinning and Galvanising

(b) Synthesis of Biodiesel

(c) Atomisation

(d) Theory of adsorption

(e) Green synthesis of indigo.

INDEX